Praise for *Enviromedics*

"Jay Lemery and Paul Auerbach have done a superb job of demonstrating how and explaining why climate change threatens everyone. They have taken an aspect understood clearly by few people and shown why it needs to be regarded as a central aspect of global health and well-being." —**Mark Plotkin**, PhD, LHD, president, Amazon Conservation Team; author of *Medicine Quest: In Search of Nature's Healing Secrets*

"*Enviromedics* is a compelling dive into the essential topic of health in a changing climate. Through a readable tour of the science and gripping examples, Lemery and Auerbach highlight the ways that shifting climate and related extreme conditions are affecting people. Whether direct or indirect, the connections are exceeding human limits for tolerance across global geographies. The authors underscore a key implication—responding effectively to the climate challenge can build vibrant societies with safety, health, and well-being at the core." —**Katharine Mach**, PhD, senior research scientist, Stanford Woods Institute for the Environment

"Many adolescents behave as if they are invincible. They soon learn that taking risks with no benefits can be future-wrecking. America and other nations can no longer behave this way; they cannot ignore ruined oceans, lands, and atmospheres, and thus, sick people. With *Enviromedics,* these experienced and accessible doctors are intervening at a moment of urgency, teaching and asserting the changes that their patients (all of us) must make. Every person, young and old, should read *Enviromedics.*" —**Richard J. Jackson**, MD, MPH, professor, Fielding School of Public Health at the University of California, Los Angeles; author of *Making Healthy Places*; host of the PBS series *Designing Healthy Communities*

"*Enviromedics* is the most important book ever written about how environmental changes affect human health. Doctors understand the effects of diet, stress, lack of exercise, and sleep on health and disease. But our health care system mostly ignores the effects of pollution and climate change on human health. Everyone should read this book, including policy makers, health care professionals, and anyone who cares about their personal health." —**Mark Hyman**, MD, director, Cleveland Clinic Center for Functional Medicine; author of *10-Day Detox Diet*

"It's difficult for many Americans, particularly in this politically charged period, to know how to think about global climate change. Often they have no contact with scientists or environmentalists, and perhaps little confidence in their views. But they know and trust their doctors and nurses, and they care about the health of their families. That is why Lemery and Auerbach's explanation of the link between climate change and personal health is such an important and useful contribution." —**Frank Loy**, former United States Under Secretary of State for Global Affairs

"Eloquent, gentle, compassionate, humble, and logical—just the sort of clear communication that all of us want from our doctors. This is an important book that, like Atul Gawande's *Being Mortal*, should be read by all members of the public, as well as the medical profession." —**Hugh Montgomery**, MD, FRCP, FRGS, FRSB, FFICM, professor of intensive care medicine; director of the Institute for Human Health and Performance, University College London

"Climate scientists speak about the impacts of climate change—rising temperatures and seas, and more frequent extreme weather events. In *Enviromedics*, Lemery and Auerbach successfully identify the ominous links between climate change and human health, including mental health. This incredibly well-researched book by subject matter experts from the health field provides great insight into why we should *all* care deeply about preventing the worst effects of climate change." —**Bill Ritter**, former Governor of Colorado

"Lemery and Auerbach have done the world a great service by providing a straightforward and understandable must-read book that clearly connects human health with an often overlooked and unavoidable essential variable—our environment!" —**Richard Carmona**, MD, MPH, FACS, 17th Surgeon General of the United States; distinguished professor of public health, University of Arizona

"*Enviromedics* is a timely and lucid account of the health effects caused by human-induced environmental change. The current and future consequences of this health crisis are widespread and monumental in their scope and scale. The world's poorest people are the most severely impacted, yet the most obviously under-addressed by anything approaching climate justice. *Enviromedics* should be required reading for anyone who has an interest in better understanding this unprecedented heath care crisis." —**David Breashears**, executive director, GlacierWorks; acclaimed filmmaker, author, and mountaineer

"*Enviromedics* reads like the plot of a disaster movie—but that is exactly what climate change is. It brings to life the human element and the reasons why we really need to worry and take immediate action. It is a must-read for global policy makers." —**Vivek Wadhwa**, distinguished fellow, Carnegie Mellon School of Engineering; syndicated columnist, *Washington Post*

"The health of humans and environmental conditions on Earth are profoundly intertwined. *Enviromedics* offers the reader a thoughtful, science-based discussion about the relationship between climate change and human health. This is a worthy conversation to be had among health care providers, and, even more importantly, among our general populations. The authors take their Hippocratic oath seriously for the benefit of society and our precious planet." —**Chad P. Dawson**, professor emeritus, State University of New York, College of Environmental Science and Forestry; editor-in-chief, *International Journal of Wilderness*; coauthor, *Wilderness Management: Stewardship and Protection of Resources and Values*

"Smoking tobacco was a scourge to human health until the U.S. Surgeon General issued a report on smoking and human health independent from global industrial interests. With *Enviromedics*, Lemery and Auerbach bring to light a considerably greater threat. They describe health hazards that are associated with global warming and the consequences for humanity. The authors are emergency physicians and experts in wilderness medicine who know better than most the impacts of nature on human welfare. This amazing book should be read by anyone who is not afraid to contemplate what we are doing and what we must do for the benefit of planet Earth." —**Hermann Brugger**, MD, head of the Institute of Mountain Emergency Medicine, EURAC research, Bolzano, Italy; president of the International Society of Mountain Medicine

"This comprehensive collection of information puts forth an easily understood critical concept. To form an opinion about climate change, it is essential to understand how environmental change could alter your health. The authors make a convincing case that climate change creates and distributes illnesses that impact the human condition. Their point is that treating these illnesses is not enough. People should treat the environment in the same way that we care for each other. If you are looking for a thorough understanding of environmental change and how it might be impacting your, your family's and your world's health, this book is a must-read." —**Dan Orr**, MS, president emeritus, Divers Alert Network Foundation; president, Academy of Underwater Arts and Sciences

Enviromedics

Enviromedics

The Impact of Climate Change on Human Health

Jay Lemery and Paul Auerbach

ROWMAN & LITTLEFIELD
Lanham • Boulder • New York • London

Published by Rowman & Littlefield
A wholly owned subsidiary of The Rowman & Littlefield Publishing Group, Inc.
4501 Forbes Boulevard, Suite 200, Lanham, Maryland 20706
www.rowman.com

Unit A, Whitacre Mews, 26-34 Stannary Street, London SE11 4AB

British Library Cataloguing in Publication Information Available

Library of Congress Cataloging-in-Publication Data
Names: Lemery, Jay, author. | Auerbach, Paul S., author.
Title: Enviromedics : the impact of climate change on human health / Jay
    Lemery and Paul Auerbach.
Description: Lanham : Rowman & Littlefield, [2017] | Includes bibliographical
    references and index.
Identifiers: LCCN 2017015582 (print) | LCCN 2017018739 (ebook) | ISBN
    9781442243194 (electronic) | ISBN 9781442243187 (cloth : alk. paper)
Subjects: | MESH: Climate Change | Environmental Health
Classification: LCC RA566 (ebook) | LCC RA566 (print) | NLM WA 30.5 | DDC
    613/.1—dc23
LC record available at https://lccn.loc.gov/2017015582

♾™ The paper used in this publication meets the minimum requirements of
American National Standard for Information Sciences—Permanence of Paper
for Printed Library Materials, ANSI/NISO Z39.48-1992.

Printed in the United States of America

JML: to my parents, Joan and John, whose loving guidance helped me to start; to my wife, Taryn, whose infinite patience helped me to finish; and to my girls, Maeve and Zada, who provide the constant inspiration for us to do better.

PSA: to my wife, Sherry; and children, Brian, Lauren, and Danny, who are committed to helping others. I couldn't be more proud of them.

Enviromedics /in'vīrō 'mediks/ *the effects, consequences, and study of the impacts of environmental change upon human health*

Contents

Part IV. Summation

Foreword

$\mathcal{T}$he connection between the environment and health dates from ancient times. Approximately 2500 years ago, Hippocrates, writing *On Airs Waters, and Places*, advises those interested in health:

> [I]n the first place to consider the seasons of the year . . . then the winds, the hot and the cold. . . . [W]hen one comes into a city . . . consider its situation, how it lies as to the winds and the rising of the sun...the waters which the inhabitants use, whether they be marshy and soft, or hard . . . and the ground, whether it be naked and deficient in water or wooded and well watered . . . and the mode in which the inhabitants live.[1]

Our ancestors believed that ill-effects of the environment were conveyed by *miasmas*, or harmful vapors from foul air, and this notion persisted into the nineteenth century. Even the name of a disease like "mal–aria" evokes the presumed "bad–air" origin of the ailment. The great sanitation movement in Europe that started in the early nineteenth century was premised on the need to protect inhabitants from noxious air produced by waste and refuse. In 1854, John Snow famously mapped the homes involved in a cholera outbreak in London and correctly traced the disease to drinking water, rather than air. Later in the century, as Louis Pasteur, Robert Koch, and other brilliant physician scientists demonstrated the bacterial causes of diseases, the conceptual hold of miasma began to abate, but even then, not completely.

In the 1890s, Sir Ronald Ross, working in India, demonstrated the cause of malaria to be parasites transmitted by a mosquito. In 1901, Walter Reed and colleagues in Cuba, acting on the theory of the Cuban physician Carlos Finlay, showed conclusively that yellow fever was likewise transmitted by a mosquito vector. Capitalizing on these discoveries, Dr. William Gorgas, who later served as the twenty-second Surgeon General of the United States Army, was able to eliminate yellow fever from Havana by controlling mosquitoes. When he was dispatched to Panama to direct sanitation and disease control for the U.S. forces building the Panama Canal, he had to contend with a commanding officer who was highly skeptical of the mosquito theory because "everyone knows yellow fever is caused by filth."[2] After the commander persuaded the Panama Commission to remove Gorgas, it was only the decision of President Theodore Roosevelt that kept him on duty to control the diseases that had defeated the earlier French effort, and so enabled completion of the canal.

Understanding of the nature and varieties of environmental threats to health advanced rapidly through the twentieth century. Early in the century, Alice Hamilton pioneered the field of industrial toxicology, investigating the physical and chemical exposures that diminished workers' health. Most of the early pioneers in radiation, including Marie Curie, died from the effects of radiation exposure. Ingestion of environmental lead in drinking water and household paint and inhalation of lead-treated gasoline vapors were found to have deleterious effects on children's intellectual development. Industrial toxins, including poisonous heavy metals and carcinogenic chemicals, became a recognized hazard in streams and waterways. Air pollution, including small particulates from fuel combustion in power plants and primitive indoor cooking, was seen to take a heavy toll in illness and premature death.

Many of these modern sources of environmental hazards share a common feature—they derive from human activity as much as or more than from nonhuman sources. Radiation exists in nature, but its concentrated forms on Earth are created by humans. Industries produce the goods that support modern life, while they spin off by-products that can harm the environment and humans. We celebrate the productivity of modern agriculture, but if the runoff of pesticides and antibiotics pollutes the water supply and encourages antimicrobial resistance, we pay a higher price than we realize for food. If fossil fuel energy production and indoor cooking make the air unsafe to breathe, that is a bad bargain. In the modern world, we have been in a constant struggle to obtain the comforts and products we want while avoiding or mitigating the adverse side effects. Balancing this tradeoff is complicated by the fact that the individuals and interests who typically stand to benefit from a polluting activity are not the same as the ones who will suffer the adverse health and other consequences.

As we advance into the twenty-first century, humanity faces an unprecedented challenge to environmental health in climate disruption due to global warming. The direct stress of heat, impact on food production, damage from extreme weather events, and extension of disease-bearing vectors are among the potential consequences for human health related to global warming. These and more are the subject of this eye-opening work by Drs. Jay Lemery and Paul Auerbach. This book is an introduction to the links among global warming, climate disruption, and human health. It is connected to contemporary concepts of personal health and planetary health, which inextricably link the impact of humans on the environment and ecological systems with the effects of the environment on human health. This book articulates a new stage of development for the field of environmental health that takes account of an unprecedented degree of environmental change caused by humans and the consequences for human health.

Global warming and climate disruption affect everyone's future. To anticipate, to prepare and to respond, *Enviromedics* beckons all of us to study, learn, think, and become truly engaged.

Harvey V. Fineberg, MD, PhD
President
Gordon and Betty Moore Foundation

Acknowledgments

$\mathcal{W}$e extend our profound gratitude to Robin Straus, whose encouragement and advocacy made this book possible. Drew FitzGerald provided creative insight and invaluable guidance from beginning to end. We also thank Kathryn Cardin, Chelsea Dymond, and Ronald Ottaviano for their contributions and editorial assistance.

Preamble: The Doctor's Approach

Any Emergency Department in America

"DOCTOR STAT TO ROOM ONE!"

The paramedics have just raced in with an elderly man who had dialed 911 from his home. He sits bolt upright on a gurney staring death in the face. Drenched in sweat, draped like a telephone pole with wires and with a tightly fitted translucent plastic mask strapped to his face trapping saliva and aerosolized mist, he's about to become moribund. His fallow body is emaciated—the toll from years of erosive lung disease. Our doctors have cared for him many times, and he gives us a knowing look through weary eyes beneath everted lower lids laden with dried-out yellow crusts. His breaths come in wheezes interspersed with loud crackling noises from the depths of his chest. It's a hot summer day, and the Air Quality Index is running deep in the purple—Very Unhealthy.

"How you feeling today, Sid?"

He's too exhausted to respond. Sid suffers from severe chronic obstructive pulmonary disease. He's a "lunger," as in lungs, of which he has precious little left, more scar tissue than elastic sponge. Flares of illness like this bring him to death's door quickly without warning. During his ER visits when he remains lucky and his body recovers enough to keep his pathophysiology from putting him in the intensive care unit, we're able to have him feeling human after a few hours, followed by a couple of days in the hospital for a tune-up. But when he waits too long to call for an ambulance, he winds up with a tube through his trachea attached to a ventilator machine, and he's swept into the labyrinth of intensive care. One of these days, he won't be able to come off the ventilator. This time, we resuscitate him by placing the tube and providing lifesaving oxygen, stabilize him by adjusting the breathing machine and giving him drugs and essential fluids, and then admit him to the ward so that he can recover and begin yet another rehabilitation.

He's a cat with eighteen lives. On the day of his discharge, I see him outside sitting in a wheelchair, face tilted toward the sun, eyes closed. He's breathing supplemental oxygen through slim nasal prongs piped through supply tubes that wrap around his head and terminate in his nostrils. Defiant, foolish, and addicted, he's dangerously smoking a Marlboro Red, which is strictly prohibited. He doesn't care. Live by the sword, die by the sword.

"Looking good, Sid!"

"Thanks, Doc," he says, not the least bit embarrassed to be caught in the act. "Tough to quit, but you know, sooner or later, I always turn around. . . ."

Patients such as Sid teach us many lessons, of which two stand out. First, although the media tend to focus on individuals and their newsworthy catastrophes, these are dwarfed by issues of population health, which increasingly are dominated by chronic conditions. To the extent that these conditions are generated or worsened by environmental changes, we have an obligation to seek remedies and improve the situation. Second, we need to maximize our survival opportunities, not challenge them relentlessly. Although some "die-hards" certainly are able to postpone the inevitable somewhat, generally they do so only through medical interventions requiring copious resources.

What are our current environmental conditions? Where are they headed? We shouldn't take a fatalistic approach and assume that patients such as Sid are "just a matter of time." It would be easy simply to attribute the fault to him and cast him aside because tobacco contributes to his disease, but why did Sid get sick on a hot day when the air was full of particulate matter? Whom do we blame on the days when children with asthma or cystic fibrosis who have tenuous lungs are forced to breathe the equivalent of cigarette smoke just because they inhale bad air from the atmosphere around them? It is not their fault. We have no one to blame but ourselves.

At the heart of medicine are the principles of prevention, that patients stricken ill rarely can achieve outcomes better than if they had never become sick in the first place. Despite all of the amazing cures and interventions we have at our disposal, we can't exercise away a lifetime of fast food and too many cocktails. Obesity leads to diabetes. Alcohol leads to liver disease. Mosquito bites lead to Zika virus infection. Everyone understands cause and effect.

So, we are going to use a straightforward approach to address a topic that is at the top of everyone's list for what will determine our future on this planet—namely, the environment. Specifically, we will frame current environmental issues in the context of their implications for human health as we understand them.

"SAVE MOTHER EARTH!"

Earth will survive, but the living creatures may perish. Please disregard for a moment any discussion of "hot-button" topics, such as the motivations imputed to environmental degradation or preservation. We concede up front that we are neither environmental experts nor economists. We are doctors on the front line who every day see medical repercussions on human beings of climate change, pollution, and the reduction of biodiversity. From our vantage point in the emergency department, we see "acute on chronic," meaning the sudden deterioration in a person's chronic medical condition because of external forces beyond his or her control. Furthermore, we have extensive wilderness experience, both as clinicians and recreationally.

Based on reading and interpreting data, we believe that our environment is changing because of certain human activities. However, we leave it to the readers to form their opinions about that. We will focus on how a changing environment impacts human health, not just in the developing world but globally. We believe that every proposed environmental situation, either maintaining the status quo or creating change, has medical implications for each of us. We will point them out.

We come with no agenda intended to provoke any specific regulation or law. Our goal is to spark curiosity, observation, discussion, and analysis. Believe what you will about whether or not there is global warming, sea-level change, air pollution, or increased frequency of harsh weather conditions. But do not be deaf or unknowing about the reasonable discussions regarding factors affecting the earth's condition and their medical impacts. Know that although socioeconomic deprivation magnifies the difficulties of certain situations, wealth and comfort afford no protection from whatever is happening to our environment and how it might affect human health.

Why should you pay attention to us? Shouldn't this advice come from a Nobel Prize–winning environmental scientist, or from a specially focused medical researcher? Perhaps. But we hope you will find our opinions and writing credible because we know medicine, because we are not sensationalizing, and most of all because we are doing this for your benefit and the benefit of future generations.

"Save Mother Earth" has been the catchphrase of the green movement for generations. However, we are not promoting sentimentality. We certainly care about polar bears and other endangered species, but our main goal is to save people. In the context of the great environmental debate, we are convinced that droughts, floods, hurricanes, toxic wastes, forest fires, severe tornados, and pollution-filled air make people sick, or worse.

Where is the medical profession in all of this? It is not yet in a leadership position. To us, that seems insufficient. Oil spills cause massive damage, followed by indignation but little sustained action. A nuclear reactor meltdown following a tsunami raises many questions, but unless the immediate environmental threat is featured in the news, it doesn't prompt many changes. Our observation is that in order to ignite and fuel the discussion to effect real changes, we need to bring it back to medical conditions that cannot be overlooked and that are with us every day. Human health is the common denominator that overlaps all boundaries of geography, science, politics, and economics. It dominates discussions of governments, social justice, and the future of mankind.

As medical professionals, we will spotlight the risk to humans if we erode the planet's limited biodiversity; seed the soil, air, and water with pollutants; and perturb climatic systems. We love wilderness medicine, and we realize that it would be difficult to contemplate wilderness medicine without any wilderness. We can grind our planet down to its nubs, but it probably will come back with a vengeance, likely because the grinders (e.g., humans) will have been removed. That is a point we will continue to emphasize. The ultimate recipient of the havoc wreaked by people on Earth is life itself, emphatically in the realm of human health.

So, here we are. We believe that no significant change in our approach to preservation of the planet will occur until influential people begin to believe that profoundly detrimental medical situations are linked to our environmental behaviors. This is no different from the reasoning we use to combat obesity, high blood pressure, tobacco use, genocide, and nuclear war. Nothing is good about any of these. We do not shirk in our responsibility to speak out; if anything, we need to intensify the dialogue. If we are falsely presumptive or in error, let experts come forward and counter our propositions. Arguments are much better than silence. If we need studies, then let's get them going.

THE BIG PICTURE

We understand that showing that a particular industrial source has an adverse effect on global air quality can be daunting. First, someone needs to measure the air for different chemicals and substances, such as sulfur dioxide, carbon dioxide (CO_2), nitrous oxide, particulate matter, ozone, organic pollutants, and so forth. Then, scientists must study various modes of transmission to ascertain the origin of the pollutants. Did the toxins come from local factories, or did they blow in from a neighboring state, or even from locations many nations distant? Are automobiles the culprits? Factories? Containerships? What portion of the pollutants is not man-made and might have originated from natural sources?

Next, researchers must measure the impact accurately. How exactly does air pollution affect people? Plants? Animals? Ecosystems? Finally, regulators and politicians must consider whether and how to reduce pollutants below harmful levels in ways that can be verified and sustained without putting undue burdens on industry or irresponsibly raising energy costs to consumers.

People approach environmental concerns from different perspectives and directions. Some focus on discrete measurements, such as parts per million of a substance in the atmosphere, whereas others gravitate to emotional rally points, such as saving an endangered animal or plant. Experts and most commentators recognize the complexity of environmental issues, including persistent organic pollutants and climate change. We hope it is logical to recognize that environmental changes might be linked to human health problems, such as declining sperm counts and increases in the incidence of certain cancers. We certainly believe that there are clear links between our environment, human exposures, and adverse health effects. By writing this book, we introduce the fusion science and our newly coined word *Enviromedics*, defined as the impact of environmental change upon human health and its related study. Our intent is to enlighten laypersons about the implications of the environment on their health. Our constituents are the patients we serve—and the ones we hope to not need to serve.

BACK TO SID

Recall Sid, the man with scarred lungs. He's damaged himself to the point that every day is a struggle for existence, and when nasty contaminated air tips him over the edge, he doesn't stand a chance. He's at the extreme, but most of us are not. Why, then, should Sid's predicament concern us and be instructive? The reason is that if we continue to foul our environment, more people on the margin of poor health may be impacted adversely. That is logical. Think about Zika virus. If the mosquitoes continue to spread and bite more people, more disease will follow. Environmental changes that warm northern latitudes will facilitate their spread.

In the same way that no health harm can ensue because of enforcing a world without cigarette smoke, no harm to our health can come from crafting a future that preserves our life-sustaining resources. Earth is our only option for existence. If we ignore it, more of us will edge closer to disaster. The metaphorical cigarette Sid smoked could be one more industrial chimney spewing smoke, one more coral reef decimated, or a steady rise in average global temperature. We urge our readers to become educated, form personal opinions, and, in turn, urge our industrial-political complex to take action.

Earth will go on, no matter what we do to it. The more pertinent question is, will we?

Part I

CLIMATE CHANGE CASCADE

$\mathcal{A}$s mentioned in the preamble, we are not environmental scientists. Like you, we will continue to formulate our opinions about what might be happening to Earth's environment based upon our interpretation of scientific data and others' opinions. However, in order for readers effectively to interpret and apply *Enviromedics* to their thinking, we are required to explain *our* understanding of what certain environmental scientists claim. At the core, if global warming is not a problem, there is no need for an explanation. If it is, then we hope you will find the following explanations reasonable enough to provoke agreement or curiosity to learn more.

Environmental change, in particular global warming, has been proclaimed the biggest global health threat of the twenty-first century.[1] This dramatic statement is in part underpinned by sometimes subtle, incremental climate changes that cumulatively contribute to the total effects of global warming. As physicians, we understand that serious diseases often progress from one small and unnoticed change in normal human physiology that leads to another change, and then another. We may not see or recognize the first change, or even the second or third, because they initially may not cause discomfort or an obvious physical alteration. For instance, a single lung cell with a genetic mutation can be the origin of lung cancer; a microscopic cluster of bacteria that walls itself off within the body evolves into fulminant tuberculosis; and an unappreciated tiny tear in the lining of a superficial vein of the lower leg can initiate and then grow into a deep venous clot that breaks off and travels to a lung to cause a life-threatening medical situation. In medicine, with serious diseases, cascading cumulative effects are the rule rather than the exception.

Apply this analogy to our environment. There are four main environmental stressors from which adverse health effects might originate: rising temperature (global warming), extreme weather events, rising sea level, and increased atmospheric carbon dioxide.[2] Frank pollution of the environment by industrial toxins poses a separate set of difficulties. These stressors cause environmental changes that might be minimal at first and then grow to become more severe. For example, sea-level rise will occur incrementally. At first, there will be beach erosion. As the elevation increases, more sand will disappear, underlying rocks will become exposed, homes will fall off cliffs, and islands submerge. If extreme weather events become more common, a few scattered thunderstorms will evolve into patterned seasons of devastating deluges followed by floods and all of the problems they create.

To understand health impacts that might be caused by environmental change, we first consider the science as put forth by experts on the topic. We begin the conversation by outlining some evidence of changes happening to our planet, then briefly explore what these changes might mean for our collective health and for certain people who will disproportionately bear the brunt of future events. Our words of caution presume that the planet is warming and being affected in other ways that might adversely affect human health.

Climate Change 101: A Primer

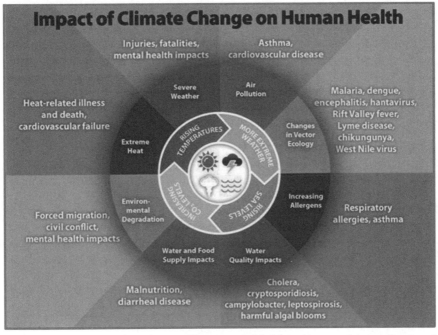

Figure 2.1. *Source:* **"CDC Climate Effects on Health," http://www.cdc.gov/climateand health/effects/, December 22, 2014. Accessed June 13, 2016.**

*W*ithin the scientific community, the most respected source that articulates climate change impacts is the Intergovernmental Panel on Climate Change (IPCC), which is convened under the auspices of the United Nations. Since 1988, it has put forth quadrennial reports that reference the most recent scientific, technical, and socioeconomic information. These reports, which are expected to be scrutinized intensively by believers and skeptics alike, are considered to be the "gold standard" by scientists on the best available climate science.[1] Between 2013 and 2014, the IPCC released its 5th Assessment Report (AR5). According to the IPCC and experts who concur with its opinion, the overall consensus on what is happening to our environment is considered by many no longer to be a matter of debate.

Former IPCC Chair Robert Watson: "Up until now, the criticism has been that climate science is like a house of cards, and if you pull out one or two sets of data, it all collapses. That narrative has been refuted. [AR5] shows that . . . the observational evidence for human-caused warming is overwhelming, compelling, and irrefutable."[2]

The report commences with assessment of the atmospheric "greenhouse gasses" that are driving increased planetary temperature. These greenhouse gasses are so named because one of their effects is to act as insulation that traps heat within our air and oceans. Concentrations of carbon dioxide, methane, and nitrous oxide have increased to levels unprecedented over the past 1.2 million years.[3] Carbon dioxide concentrations in particular have increased by 40 percent since the mid-nineteenth century.

The science done to determine this fact is ingenious. Measurements are taken from ice cores drilled deeply into the glaciated regions of Greenland, Antarctica, and North America.[4] Much like the rings seen on cross-sections of cut trees, the ice samples allow determination of the events and time line of ice and snow deposition that have taken place over eons. As snowfall accumulated in the polar regions, air became trapped within each successive layer of ice and snow. Frozen in time, the trapped air serves as reliable atmospheric samples from the moments of capture. Thus, ice core sampling and analysis have provided scientists the ability to measure our atmosphere, notably carbon dioxide and methane, as it existed at discrete intervals over eons.

Data from the ice cores are disturbing. Experts note exponential increases in air concentrations of carbon dioxide and methane over the past 150 years to levels never seen previously. In 2013, atmospheric levels of carbon dioxide exceeded 400 parts per million (ppm) for the first time ever. These levels were measured at the Mauna Loa Observatory on the "Big Island" of Hawaii. By comparison, the historical average prior to the year 1850 was 270 ppm.[5]

What does this all mean? Greenhouse gasses trap heat within the atmosphere and oceans. AR5 reported that each of the past three decades has been successively warmer than any decade before 1850 for which temperature was recorded. In the Northern Hemisphere, the years 1983 to 2012 were likely the warmest thirty-year period of the past fourteen hundred years, and 2016 was the warmest year in the history of weather record keeping.[6] As air temperatures rise,

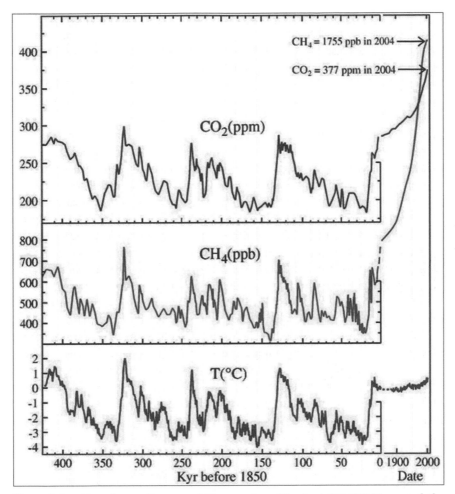

Figure 2.2. Greenhouse Gasses and Temperature: The Last 500,000 Years and the Last 200 Years. *Source:* NOAA Centers for Environmental Information, https://www.ncdc.noaa.gov/data-access/paleoclimatology-data/datasets/ice-core. Accessed June 17, 2016.

Notes: CO_2=carbon dioxide; CH_4=methane; ppb=parts per billion; ppm=parts per million; T=temperature; °C=degrees Celsius; Kyr=thousands of years

oceans also warm, contributing to shrinking and thinning of the Arctic Sea ice cover, reduction in the volumes of glaciers worldwide, and decrease in Northern Hemisphere spring snow cover. Regarding the impact on snow cover, because the dark-colored surface of land and water reflects less sunlight than does light-colored snow, warming is compounded by the *albedo refractive effect*. In other words, dark tundra and ocean absorb more sunlight than does the reflective snow, accelerating energy uptake by, and warming of, these darker surfaces.

A warmer planet directly threatens planetary glaciers, particularly those of Greenland and Antarctica, where 2.1 percent of all the water on Earth exists in a frozen state.[7] As glaciers melt, they increase the amount of water in the oceans. According to AR5, at the current rate of glacier melting and water transfer to the oceans, sea levels globally could rise between three and five feet by the end of the twenty-first century. We already have seen evidence of this. During the period 1901 to 2010, global mean sea level rose by 7.4 inches, a rise that accelerated between 1993 and 2010.[8] Aggregate predictions are that over the next hundred years, sea-level rise in Ocean City, Maryland, might reach numbers like 2 feet; Guam, 3 feet; and Manila in the Philippines, 4 feet.

Points of view differ about whether or not rising carbon dioxide levels in the atmosphere are categorically harmful. Some persons believe that modest global warming (approximately 1 degree Celsius [1°C], or 1.8 degrees Fahrenheit [1.8°F]) over the next century attributed to carbon dioxide rise might be beneficial if it improves agricultural yields and therefore food security.[9] This point of view must be counterposed against the argument that this temperature rise will then be surpassed and contribute to the bad effects of global warming. As we have already mentioned, we are writing under the assumption that human modification of the environment has the potential to cause ill health, which must be taken into account in any reasonable discussion about what should or shouldn't be done to deal with global climate change.

Greenhouse gasses do not just affect temperature of oceans, but also their chemical composition. Seawater has absorbed the majority of global carbon dioxide increase, which has directly increased the water's acidity (i.e., lowered its pH). Acidification of oceans alters the biochemical processes of marine life and has been implicated in the collapse of major fisheries and disruption and "bleaching" of the biodiverse coral reefs, both effects quite detrimental to human food security.[10]

We understand exceptions to trends are almost certain. For instance, changes in the global water cycle will not be predicted with absolute certainty or always be uniform. If our weather is changing, the contrasts in precipitation between wet and dry regions and seasons will increase, yet there may be regional exceptions. Complex systems can sometimes be affected so that they respond in ways that are not predicted. Although an increase in global warming has been noted throughout the world, including July 2016 being the hottest month ever recorded, the northeast United States witnessed one of its coldest and snowiest winters on record in 2015.[11] That might be attributable in part to global warming and its effects on air circulation patterns. In the face of all of these data and theories, we should keep in mind that it is the trend that matters, to the extent that it can be determined and predicted. Sick patients have good days, but they are still sick. Such is the challenge of science communication in the face of seemingly contradictory data.

"Climate change" and "global warming" are the most commonly used descriptors that have gained acceptance as terminology for our current situation. Perhaps a more accurate description is *climate energizing*. Rather than thinking about global warming purely as a "hotter" planet, consider what is happening: human activities are adding energy to a very complex system. It is too simple to conceive of our planet as a mere ball of dirt, whereby adding heat to it will merely create a warmer ball of dirt. Earth and its ecosystems form a dynamic entity with a practically infinite number of moving parts. So, adding energy does not just make the system warmer, but makes it more *unpredictable*. Heat is energy. As more energy is added to the system, there is more "fuel" for storms, as observed in tropical cyclone intensity and storm surges. Recently we witnessed the two most powerful storms in recorded history: Typhoon Haiyan (2013) generated maximum sustained winds of 196 mph, and Hurricane Patricia (2015) maximum sustained winds of 200 mph.[12,13]

Unpredictability about whether these changes will be unrelenting is still a common feature of the unknowns of climate change, but it is considered to be highly likely for any understanding of environmental changes that might undermine human health. People have never been through anything approaching our current situation of industrialization, intrusion upon the environment, and perhaps global climate change. From a medical predictive perspective, it is totally reasonable to assume that climate change is occurring.

Most of us are familiar with the complex system represented by a three-year-old toddler. How would this system respond to an input of energy in the form of a sugar-laden candy bar? We would see a hyperactive child who predictably "bounces off the walls." But there is a good chance that the dynamism might turn into unpredictable behaviors: aggressiveness, mania, or even extreme fatigue. For the child, the erratic behaviors are considered outliers, well outside his or her historical behavior. This is the essence of what many observe about climate change—an energized world that exhibits increasingly unpredictable behavior. This is how it will be until climate and weather patterns become regular. By the time this happens, we should not be surprised to see adverse impacts upon human health.

We next introduce general concepts and assertions related to this notion of cause and effect.

LINKING CLIMATE CHANGE AND HUMAN HEALTH

Certain predictions are that until approximately 2050, climate change predominantly will impact human health by worsening preexisting problems in people, thereby acting as a *threat multiplier*.[14,15] For instance, a person with asthma might have more frequent or more severe asthma "attacks" because of air pollution. Some predict that after 2050, climate change will lead to increases in the baseline number of persons with ill health.[16] The increases will be due to exposure to more intense weather and temperature conditions, a greater local number of potentially hazardous conditions, and longer duration of harmful exposures. Residents of low-income ("developing") countries will be affected disproportionately because they lack certain protections afforded to persons residing within privileged, wealthier nations.[17]

Extreme weather will both cause direct injury and act as a threat multiplier to the most medically vulnerable. Weather patterns can be considered as residing under a standard bell-shaped curve. This is a statistical model shaped like a bell, which is used to show the distribution of common events, as well as the likelihood of uncommon events ("outliers"). In any bell-shaped curve representation of weather, most events happen with the frequencies noted under the big part of the bell; rare events (such as extreme temperatures, droughts, heavy precipitation, high winds) occur with the frequencies represented by the tail ends of the bell, which are generally far less than the common events. This makes great sense as we observe our current world weather situation.

But what about predictions for the future? What appears to be happening with climate change is that the bell-shaped curve that represents weather patterns is both moving to the right (i.e., warmer temperatures) and becom-

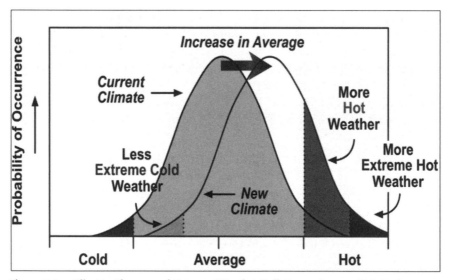

Figure 2.3.   Climate Change and Extreme Weather Bell Curve. *Source:* U.S. Environmental Protection Agency (EPA), https://www.epa.gov/climate-change-science/understanding-link-between-climate-change-and-extreme-weather. Accessed October 4, 2016.

ing flattened, which means that extreme events are becoming more common. Among these uncommon events, some are far more intense than any in historical memory.

One example of the influence of socioeconomics upon vulnerability was the European heat wave of 2003. Even though northern Europe overall was relatively well resourced, poor people still suffered. These countries had little experience with what to do when temperatures exceeded 38°C (100°F) for a week straight.[18] Apartments originally constructed in the 1800s had no central air conditioning. People did not know until too late that they needed to check on their elderly neighbors, and public health officials did not realize that shelters were a public health necessity to provide lifesaving access to water and air conditioning. Twelve thousand persons perished in Paris alone, mostly in the poor neighborhoods.[19]

As we witness a greater number and intensity of extreme weather events, we can expect similar outcomes correlated to socioeconomic conditions. The influence of societal class, preexisting illnesses, and age will be repeated throughout this book. As is also commonly the case with certain diseases, the most vulnerable—the poor, sick, and elderly—will bear the brunt of the impacts of climate change with respect to adverse health effects.

Extreme heat more severely impacts persons with preexisting physiologic vulnerabilities, such as those with heart failure. This leads to increased hospitalization or even death. A study published by researchers at the Harvard School of Public Health clarified that for each 1°C (1.8°F) increase in summer temperature, the death rates for elders with chronic medical conditions, such as diabetes, heart disease, and chronic lung disease, increased by an amount of 2.8 to 4 percent.[20]

JUST A FEELING

The veterans of fighting the fires could sense when they were coming. Some might call it instinct—others might consider it experience—but the vets were nearly infallible. Hot, dry weather in the Pacific Northwest was an anomaly when they were youths, but now it was the norm. Entire summers went by without so much as a raindrop. Millions of acres of dense forest that had once been wet, moss-covered, and spongy underfoot were crackling and crunchy. The dark green filaments that hung like a troll's beard on the limbs and trunks of ancient ponderosa pine had been transformed into yellow and brown threads of tinder waiting to explode in the next conflagration. Thunderstorms didn't bring relief. Their lightning started the fires, and there was hardly enough rain to dampen a community park, let alone an expanse of wilderness.

Craig Wells had seen it all. He'd fought fires in Washington, Oregon, Idaho, and Montana. He'd jumped into their midst from airplanes, dug fire lines, and intentionally burned areas to keep advancing fires from eating up the landscape. All of that smoke, the lousy Air Quality Index, and a nasty cigarette habit he attributed to stress relief had scarred his lungs beyond repair. On most days now, an inversion layer in the valley trapped the pollution so that he became a prisoner in his own home. Without the filters on his air conditioner and an occasional afternoon breathing oxygen through nasal prongs from a tank he wheeled beside him, he was breathless. And when the fires came, it all became worse. His kids wanted him to move, but it was pretty much the same everywhere he might live, unless he was willing to move to the desert or North Pole. He couldn't afford to relocate. He grew up here, and by God, he would die here.

That morning, he awoke with "a feeling." Just a spark would do it. After forty-five consecutive summer days without so much as a cloud in the sky, the anvil cloud in the distance warned of a disaster waiting to happen. By lunchtime, there was a big wind. Not good. In the afternoon, he heard rumblings of thunder in the distance. There was nothing anybody could do. When he heard the second helicopter fly over his home in the woods and saw sparks crowning the treetops, he knew the worst was yet to come. Like he had done so many times before, he packed the car with his most precious belongings, hosed down the roof, and waited to be ordered to evacuate.

What brought him down were the shifting winds. Driven by colliding weather fronts and the whims of nature, the Wenatchee River Fire, as it came to be called, was a merciless beast. It laid down a tidal wave of smoke that came so fast that it couldn't be evaded. No flames. Just smoke. Craig went inside and did everything right—shut the windows, sealed the openings, put a moist cloth over his nose and mouth—but he suffered irreparable harm. His scarred airways were no match for the invasion of carbonaceous particulate matter. He felt his chest tighten, and he struggled to breathe. He couldn't drive, because he was too weak and couldn't see. His inhaler didn't help at all, so he dialed 911 and prayed that the ambulance could beat the fire to his home. A dozen calls came in to the dispatch center at exactly the same time, all asking for assistance. They came for Craig, but by the time help arrived, it was too late.

Another force at work that contributes to vulnerability during extreme heat events is urbanization. Cities and climate are coevolving in a manner that will amplify the effects of heat and vulnerability of urban populations to heat-related illnesses. More than half of the people on Earth now live in cities, up from 30 percent fifty years ago.[21] By 2030, approximately 60 percent of the projected global population of 8.3 billion people will live in cities.[22] It is predicted that population growth will be accompanied by rapid urbanization, which will add to cities' thermal storage capacity, resulting in an *urban heat island effect*.[23] Because of this phenomenon, urban heat islands can add 3.5°C (6.3°F) to the ambient temperature relative to immediately adjacent rural areas.[24] This additional bump in temperature has been linked to increased mortality. Studies have shown that lower socioeconomic and ethnic minority groups are more likely to live in warmer neighborhoods with less open park space. This places them in an environment where they must endure greater exposure to heat stress with limited additional coping resources, such as air conditioning, swimming pools, or reflective roofing.[25]

It is well known that in the fall of 2012, at its peak, Superstorm Sandy covered more than one thousand miles of North America's coastline.[26] But a lesser-known story of Sandy speaks to the proposition that vulnerable populations suffer disproportionately. Bellevue Hospital, a New York City institution since the eighteenth century, is the public hospital flagship of the city's system, dedicated to treating the poor and underserved. Prior to Hurricane Sandy, it had never closed. The storm surge from Sandy caused a catastrophic power failure and changed all that. To achieve repairs to damage the hurricane caused, the hospital had to close its doors for nearly a year. Bellevue was not alone; just eight blocks away, the NYU Langone Medical Center also was knocked off-line by storm surge from the East River. Although other hospitals throughout Manhattan were able to absorb patients on an emergency basis, hundreds of thousands of patients lost access to their regular doctors, treatments, and medicines. Persons of means did not suffer anything approaching this kind of discontinuity in their medical care. The cautionary tale of Sandy is that even in one of the most resource-rich cities in the world, an extreme weather event was able to exploit unforeseen vulnerabilities and interrupt health care for a very large number of its citizens.

Extreme weather sometimes causes flooding. One notion in the scientific field of microbiology views the world as being covered by a "fecal veneer," in which our planet is blanketed by layers of infectious filth.[27] Public health efforts have focused on keeping this veneer thin where people live, trying to separate humans from sewage and agricultural fields (containing livestock and fertilizer), where the layers are thick. If these efforts are effective, we are separated from germs that contaminate our food and water and cause disease. However, heavy rains overturn all of this protection by flooding farmlands and causing landslides, which overwhelm sewage systems, contaminate drinking water, and spread infectious diseases. Churning up the landscape and flooding it creates a blender effect, mixing the fecal veneer with the same water that is used to sustain hygiene, hydration, and nutrition. When heavy rainfall combines with increased water temperatures and agricultural nutrient runoff, algal blooms can arise and grow with amazing speed. These contaminate fisheries and exude toxins that are poisonous to humans in many ways, including liver and lung inflammation and neurologic disease.

Weather extremes highlight the dramatic face of climate change because they are events that are impossible to ignore and cause quite obvious medical sequelae. But slow, insidious processes related to climate change are far less obvious, yet potentially more threatening. One of the most worrisome of these processes is the spread of vector-borne diseases, such as malaria and dengue. Disease-carrying mosquitoes have responded to warmer global temperatures by expanding their habitats to more temperate regions and higher elevations. After its previous eradication in southern Europe, malaria has reemerged in that location. Dengue viral disease occurred in Portugal in 2012 for the first time in nearly a century.[28] There is a large increase in mosquito habitat in the mountainous regions of East Africa.[29] There is also an unprecedented occurrence of *Plasmodium* malaria transmission in Alaskan resident and migratory bird species.[30]

Global warming will cause losses of biodiversity in temperate and tropical regions of the world.[31] Warming temperatures and unpredictable weather also will cause net decreases in agricultural yields and undermine food security and human nutrition.[32] Most developing nations, many of which are very poor and some arguably overpopulated, are in the equatorial belt, where the greatest loss of agricultural productivity will occur.[33]

Even worse off will be human inhabitants who literally lose the land under their feet. It is predicted that by the year 2100, sea-level rise will force displacement of up to fifty million citizens of low-lying island nations.[34] The physical and emotional stresses of forced migration are profound and life altering. Large populations will be uprooted, separating families and friends, and disrupting commerce and the ability to live and work in a familiar location. The health risks that environmental refugees will endure will be numerous and long lasting.[35]

Now let us consider in more detail some of the health impacts of global climate change.

· 3 ·

Heat Waves and Heat Stress

$\mathcal{G}$lobal warming means that Earth's ambient environmental temperature is rising. In the most obvious causal relationship, warming equals heat. It is getting hotter outside. Numerous consequences, many of them discussed in this book, include changing weather patterns, melting polar ice, rising sea levels, and northward migration of disease-carrying mosquitoes. However, the hot environmental temperature alone poses a direct threat to humans, whose bodies have not evolved to naturally withstand this amount of heat.

Warmer temperature trends are sometimes measured in fractions of a degree evolving over intervals of years. Interspersed with these trends are more pronounced spikes in temperatures measured in tens of degrees that last for days or weeks. Thus arise heat waves, which are extreme climatic events that put people at immediate risk for physical suffering and rapid-onset, catastrophic medical derangements. When outside temperature soars during a heat wave, it can become truly horrific from a public health perspective. The effects of heat waves are very well understood. Nothing speaks more directly to the global impact of planetary warming on human health than what happens to a person who is exposed to extreme environmental heat. In such a broad exposure across large populations, it's not uncommon to have thousands of people harmed or killed by rising temperatures and extreme weather perturbations.

TOO HOT

Philadelphia can be intolerably hot and humid in the summer. "Muggy" means more like getting mugged by the heat. Walk down Broad Street near Temple University and watch the shimmering waves of hot air rise from the pavement.

19

Step over the bubbles of sticky asphalt, or the bottom of your shoes will turn black and nasty.

Daisy's apartment was right across the street from the hospital, but she was too old and proud to leave her cramped, stifling quarters and make her way to sit in the air-conditioned lobby. Last Thanksgiving, when it seemed that Daisy would be alone for the holidays, the landlord had deposited her in the waiting room of the ER, knowing that the doctors would admit her to the hospital, assuring that she would be fed and cared for over the holidays. But now Daisy was resolute, and she "wasn't going nowhere." From now on, she was determined to be on her own, come hell or high water. Old folks can be stubborn, and Daisy was no exception.

Daisy didn't remember the summers of her youth as being so hot for so long. In the old days she could go to the beach, play in the sprinklers, and drink lemonade in the shade. When you're ninety-three and living alone, you just sit and swelter. You can't do anything quickly. Her kids could have been thousands of miles away, for all they paid attention to her, and the landlady was irresponsible, so the fans and air conditioning seldom worked.

After the temperature records were broken for ten straight days, 90-plus turned to 100 and it became really dangerous. The electrical grid was overwhelmed, and the first brownout turned off 100,000 cooling units. Daisy's was one of the first to fail. She simmered, yet paradoxically, her thirst disappeared. It was nauseating to drink tepid tap water. So, she stopped taking her heart medication, and her ankles began to swell. The demands on her body were too great. As Daisy's body tried to cool, her heart rate increased, and the tired pump muscle began to fail. We take a lot for granted when we're young, including being able to survive a heat wave. That doesn't apply to older people.

She didn't stand a chance. Daisy suffered like the others in the stifling heat, and her old heart and nervous system couldn't handle the enormous stress to her system. Too tired to leave, she sat on the couch and sweated until she had no more sweat to give, and became more dehydrated and tired while her organs began to fail. After her kidneys shut down, she slipped into unconsciousness. Even if she had awakened, she wouldn't have remembered, because a stroke from a leaky artery in her brain wiped out her memory.

Could this all have been avoided? Well, Daisy wasn't going to get any younger, but she didn't need to die when she did. In the City of Brotherly Love, the heat wave claimed fifty victims in a week, most of them senior citizens who simply didn't have the knowledge, resources, or ability to cool themselves before their bodies called it quits. Across the nation, in the same week, hundreds more succumbed to the heat. With a warmer planet, this would undoubtedly begin to happen every summer to increasing numbers of victims.

HEAT-RELATED ILLNESS

Because heat-related effects on humans rhetorically are the most obvious direct consequence of global warming, we want to explain this relationship further. The following is adapted from the sixth edition of *Medicine for the Outdoors*:

Human core temperature is maintained at 37°C (98.6°F), with little variation from individual to individual. Heat is generated by all of the activities that contribute to life, from the blink of an eyelid to the completion of a marathon. If it is not shed constantly at approximately the same rate as it accumulates, a person overheats. As we learned in high school science class, heat is lost to the environment through conduction (heat exchange between two surfaces), convection (heat transferred from a surface to a gas or liquid), radiation (heat transfer between the body and the environment by electromagnetic waves), and evaporation (consumption of heat energy as liquid is converted to a gas).

Humans evolved to tolerate a certain climate, and in general fare poorly in hot environmental temperatures. In the heat, extreme humidity impedes evaporation and greatly diminishes human temperature control. The National Weather Service heat index serves to guide us:

Table 3.1.   The National Weather Service Heat Index

Apparent Temperature Range	Dangers/Precautions at This Range
27°C to 32°C (80°F to 90°F)	Exercise can be difficult; enforce rest and hydration
32°C to 41°C (90°F to 105°F)	Heat cramps and exhaustion; be extremely cautious; provide constant supervision
41°C to 54°C (105°F to 130°F)	Anticipate heat exhaustion; strictly limit activities
54°C and above (130°F and above)	Setting for heatstroke; seek cool shelter

Source: National Weather Service, http://www.nws.noaa.gov/om/heat/heat_index.shtml. Accessed October 4, 2016.

When heat-control mechanisms are overloaded, the body responds unfavorably. The syndromes of true hyperthermia (in which core body temperature is measurably elevated) can rapidly become life threatening as the heat destroys vital organs and dismembers chemical systems essential to life. The body may lose its ability to control its own temperature when it reaches 41.1°C (106°F), so from that point upward, body temperature can skyrocket. Heatstroke is undeniably a profound medical emergency. When the body's enzymes begin to denature, in essence being "cooked," these vital chemical messengers that maintain our metabolic processes cease to function, liver and kidney failure follow, and the brain and central nervous system malfunction prior to shutdown. Extreme heat kills.

Heat exhaustion is illness caused by an elevation of body temperature that does not result in permanent damage; heatstroke can permanently disable a person. The signs and symptoms of heatstroke are extreme confusion, weakness, dizziness, unconsciousness, low blood pressure or physiological shock, seizures, increased bleeding (bruising, vomiting blood, bloody urine), diarrhea, vomiting, shortness of breath, red skin rash, and tea-colored to deeply darkened ("machine oil") urine (caused by breakdown of muscle tissue that releases pigment into the bloodstream).

In addition to heatstroke, rising environmental temperatures can be expected to cause more episodes of muscle cramps, swollen feet and ankles ("heat swelling"), heat exhaustion (which immediately precedes heatstroke), and fainting. The very young and old will be affected disproportionately by the heat, in part because they cannot protect themselves and because their bodies do not regulate body temperature very well. Hospitalizations for heat-related illnesses will rise, and fatalities will increase. Persons without air conditioning can be expected to suffer more than those who have cooling units.

The geriatric population has a limited capacity to self-advocate and adapt, as well as to acquire critical resources: water, cooling, and medical attention. Such was the case with many of the victims of the 2003 European heat wave, who had little knowledge of the essential nature of hydration and lived in dwellings with stifling airflow, without fans or air conditioning. Their apartments became lethal saunas. The end result was a physiologic catastrophe for thousands and a continent-wide trauma that struck ill-prepared nations.[1]

At times it can be difficult to prove that a particular event is directly caused by climate change. But in the case of the 2003 heat wave disaster, anthropogenic greenhouse-gas emissions made such an event far more likely than it would have been in the absence of such changes.[2] Furthermore, the European heat wave fell within global circulation climate models projected for the early twenty-first century. With current emissions rates, the likelihood of such events will increase approximately a hundredfold by 2040.[3] By comparison, the summer of 2003 will seem very cool.

The "all-cause mortality" approach, in which correlations between environmental heat and causes of death are revealed, shows several relationships. Above a certain temperature threshold, mortality (injury or illness) rises side by side with temperature. These causes of disability and death are strokes, heart attacks, respiratory failure, and heatstroke. Heat stress strongly correlates with exacerbations of lung disease, such as asthma and emphysema, as well as dehydration and injurious electrolyte abnormalities. It's estimated that each approximately 1°C (1.8°F) increase in summer temperature increases the death rate for elders with chronic conditions by 2.8 to 4.0 percent.[4] Further stratified, the approximate estimated death rate increase per 1°C (1.8°F) increase in temperature is:

- 4.0 percent for persons with diabetes
- 3.8 percent for persons with a previous heart attack
- 3.7 percent for persons with chronic lung disease
- 2.8 percent for persons with heart failure

There are other ominous trends. During heat waves, concrete, blacktop, and the low ventilation capacity of urban "canyons" created by tall buildings can add 3.9 to 6.6°C (7 to 12°F) to the urban heat load.[5] This phenomenon explains why cities are hotter and retain heat longer than do nearby rural areas. Today, more than 50 percent of the world's population lives in cities (compared with 30 percent in 1950).[6] This percentage is increasing, and in future years it will place more people under the "urban heat island effect." To compound this susceptibility, by 2100 (relative to 2000) there will be one hundred million more people alive over the age of sixty-five years.[7] As rising environmental temperature intersects an aging population, the risks for medical problems, particularly exacerbation of chronic diseases, become obvious.

WELCOME TO HELL: THE MIDDLE EAST 2050

Sunny was thrilled to have landed a construction job in Qatar. His salary would go a long way to pulling his family out of squalid living conditions back in the slums of Lahore, Pakistan. He considered himself lucky.

Heat and humidity were constant in the Persian Gulf. The legendary heat had always made it an excruciating place to work, but over the past few decades, the weather had come to resemble one of Dante's Circles of Hell. Sunny was intimately familiar with the local risks—one of his uncles had perished in the push to complete the World Cup stadiums in 2022. The official meteorologic cause was a heat wave.

Extreme weather events were now the norm. The upcoming week's forecast for Doha was predicted to exceed a heat index of 74°C (165°F) with a "wet globe bulb" reading of 35°C (95°F). Still, Sunny was undeterred. The construction project was behind schedule, and the bosses were offering double salary to work though the hottest noon hours. No problem, he confidently mused as he carried a gallon of water to the site for an 8 a.m. start.

The wet globe bulb measures both temperature and humidity using a thermometer wrapped in a moistened cloth. It's how heat index is determined. This measure of ambient heat stress upon the human body is considered to be more predictive of heat illness than environmental temperature alone. On this day, the heat index would outpace Sunny's ability to cool himself, regardless of hydration and an occasional break from exercise. Sweat cools the body by evaporating into the surrounding air, taking body heat with it. In the stifling conditions of a Middle East extreme heat event, the air is so full of moisture that sweat evaporation cannot occur, leaving the body to slowly steam and roast.

Twenty minutes into the workday, Sunny was drenched with sweat. He drank continuously and tried to seek the shade, thinking this would help him beat the heat. His body temperature rose. By 11 a.m. he had a throbbing headache and felt weak and sick to his stomach. By noon he had guzzled so much water that he had diluted the sodium concentration in his body to a dangerously low level. By 1 p.m., he was repetitively vomiting and so confused that he believed he was talking to his grandmother in Pakistan.

He collapsed and was rushed to Hamad Hospital, where the astute doctors immediately immersed him into an ice water bath. His core body temperature was 42°C (108°F), hot enough to kill him. But he was lucky, because the doctors cooled him in the nick of time. After a rocky week in the intensive care unit, he eventually made a full recovery. Such was the risk for laborers in the Middle East.

A NEW EPIDEMIC . . .

Life had never been easy for a farmer in El Salvador, but compared to the murderous years of civil war, hard labor in the heat seemed like an improvement. Mauricio was grateful to raise his family in a time of peace. Sugarcane was in high demand, and the work was steady and reliable. After twenty years in the field, he thought he knew how to keep himself safe.

It was punishing labor, and over the years the tests of endurance had ratcheted upward. Daytime temperatures grew hotter, and drinking water more scarce, either dried up or diverted for agriculture. Like most of his colleagues, Mauricio was suspicious of the local water source in the cane fields, because it was brackish and certain to give a man diarrhea. It was his daily custom to hydrate with a liter of Fanta and boiled coffee.

At thirty-six years old, the formerly rugged farmer had a reputation for being one of the strongest workers, but over the past few months, his body was changing. He was losing his stamina. By midday, he was winded and couldn't keep up. His legs were visibly swollen. Waves of nausea were a daily routine, and he suffered from mal de orín, peeing painful sand-like material, sometimes with the appearance of rust.

Sunday was his day of rest, and today he spent all of it in bed, disappointing his five children, who implored him to play. Before the new week began, he told his wife that it was time to make the six-hour trip to Hospital Rosales in San Salvador.

Within minutes of hearing Mauricio's story, the doctor sighed. He had heard it many times before, and he was fearful. Later that afternoon, when the blood tests confirmed his suspicion, he broke the sad news to his patient. Mauricio's kidneys were shutting down as he became another of the many young victims of Mesoamerican nephropathy, an emerging epidemic of kidney failure caused by prolonged exposure to extreme heat and lack of consistent access to potable water.

There was no cure. If he had the financial resources, Mauricio could spend three mornings per week connected to a dialysis machine, but that was an impossible proposition for a sugarcane worker who lived hours away from a medical center. And for a poor man, a kidney transplant was all but impossible in this part of the world.

INTO THE FRYING PAN . . .

A hotter climate will not affect all humans equally. A disproportionate burden will be borne by the most vulnerable: laborers who have no choice but to work outside in the heat; lower socioeconomic communities living in urban islands of rising temperatures; and the elderly, whose chronic medical conditions will cause them to unravel from unrelenting physiological stress caused by exposure to hot climates, whether indoors or outdoors.

Extreme heat is already the leading cause of weather-related deaths in the United States.[8] During our summers, more people succumb to hot air temperatures than to flooding, lightning strikes, tornados, and hurricanes combined. As average temperatures rise and extreme heat waves expand their temporal footprint to occur in the late spring and early autumn, our traditions of summer picnics and visits to the beach will be invaded more often by public health "red alerts" and emergency cooling shelters. This may not seem like such a burden now, but when it begins to happen, we may wish we had done more while we still had time.

• *4* •

Extreme Weather

$\mathcal{E}$arlier we introduced the concept of "climate energizing"—that a warmer world doesn't just mean higher temperatures, but that more energy is being added to weather systems. The result is frequently extreme and unpredictable weather (both low-precipitation and high-precipitation events) that will broaden the boundaries of historic weather patterns for a given region. Thus, droughts, floods, and fires that once would have been considered unusual events will become the new normal. Hyper-energized weather now tests the capacity and reserves of ecosystems throughout the world to absorb (and sometimes, even to survive) these more intense events. Human resilience is clearly tested, so that global increases in illnesses and injuries caused by extreme weather likely constitute parallel concerns.

According to the United Nations Office for Disaster Risk Reduction, weather-related disasters in the past two decades have killed more than six hundred thousand people and inflicted economic losses estimated at trillions of dollars.[1] Another 4.1 billion people have been left wounded, displaced, or otherwise in need of emergency assistance.[2] When considering significant weather-related disasters (defined as having been responsible for more than ten deaths or having affected more than one thousand persons), the world has averaged 335 such disasters every year over the past two decades, which is double the average number from the previous ten years.[3]

Evidence is strong that a warming climate creates more frequent and intense heat waves, causing heavier rainstorms, worsening coastal flooding, and intensifying droughts. Floods account for approximately half of weather-related disasters, affecting 2.3 billion people worldwide over the past twenty years, but droughts are the biggest killers, accounting for 59.6 percent of extreme weather-related deaths.[4] Because each extreme weather pattern has unique characteristics and health effects, we address them individually.

DROUGHT

How hot is hot? How dry is dry? The "Dust Bowl" that devastated the center of the United States in the 1930s has been attributed to anomalous farming methods, the sequel to a land grab that was followed by stripping away the grasses that secured the topsoil. But it was more than that. It was marked by precious little rain, and even if the grasses had remained, the hot, dry conditions would likely have forced men, women, and children to migrate to places that could better sustain life.

Droughts dry up the earth and eliminate essential elements that support life. Without water, animal and plant food cannot be created or sustained, and what could be made fertile becomes barren. If the climate does not deliver rainfall and fill reservoirs, water cannot be distributed and crops suffer. Without water, animals raised to feed humans perish. People without water and food suffer dehydration and starvation.

We can realistically view our current situation and contemplate a future. Here's how it might read thirty years hence:

The first dry season should have been a cause for concern, but it was not. The farmers had heard tales of years like this and so were not alarmed. They didn't rotate crops, attempt to conserve water, or store extra food. They erroneously counted on everything returning to normal the next year. But it didn't. The first year's sun-parched fields looked the same the next year, and the next, and the one after that. They had no choice but to move. But, where? Global conflict and acts of terrorism long ago had caused an abrupt halt to safe mass migration, or even to seeking asylum in desperate times. With endless sunshine, scant rainfall, and effective incarceration in their parched regions, entire populations of hundreds of thousands and sometimes millions of people were held hostage to the mercy of whatever charity was directed to them by relief organizations.

Children suffer the most, because they are at the mercy of stronger adults, including their parents. In the face of self-preservation, adults can defend themselves slightly longer. What chance does a two-year-old child have when she cannot be fed from a shriveled breast and cannot compete for a ladle of fetid, algae-infested water from a barrel into which she cannot reach? When the tears no longer come because there isn't enough moisture in the body to create them, the days of life are numbered. Sunken eyes and a mute affect cannot express the misery of dying from dehydration and starvation. The most severely afflicted have no strength and thus no voice. Their cries are silent.

Dramatic disparities exist. While stadiums filled with fed-and-watered sports fans celebrate another victory on the Olympic playing field on the other side of the world, entire populations are crowded into camps without water and food, and so these forgotten people promptly wither, die, and are forgotten.

Droughts are defined as prolonged periods of abnormally low rainfall. It's predicted that they will become more frequent and intense in coming decades.[5] The health effects of drought accumulate over time, and beyond acute starvation are mostly indirect, meaning that drought optimizes conditions for other health threats, such as infectious diseases or chronic malnutrition. The health impacts of a drought depend on its severity, baseline vulnerability of the population, state of public health infrastructure, and community resources available to absorb the drought's effects. People may die from dehydration during a drought, particularly when they are stranded. More commonly, deaths occur from famine, malnutrition, waterborne infectious diseases, social strife, and heat-related illnesses. Socioeconomic factors (sometimes called the social determinants of health) are significant cofactors, as we will discuss.

The most widespread health impact from drought is malnutrition, with the biggest burden afflicting low- and middle-income countries. Malnutrition is the endpoint of a much larger concept of *food security*, which describes the multifaceted process of moving food from the fields (crops and livestock) to the dining table. Drought stresses ecosystems and hinders available food supplies, reducing the quantity and/or quality of nutrients. Malnutrition follows and degrades a person's overall health, leading to greater vulnerability to illness and disease.[6] Abundant scientific evidence supports the prevalence during droughts of malnutrition and disease, with the most common being diarrhea, pneumonia, and other infectious diseases.[7,8]

In places with tenuous food security that are overly dependent upon a staple crop, major diet changes will result during crop scarcity. The resulting adaptive behaviors may include consuming unfamiliar foods or food that is improperly processed. Recent history demonstrates what happens next. In Afghanistan in the early 1970s, liver disease affected 22 percent of a population that had consumed wheat that had not been properly separated from a plant contaminant. Known as Charmak disease (or "Camel Belly"), this potentially fatal liver disease is caused by eating wheat flour contaminated with seeds of charmak weeds (*Heliotropium* plants), which contain pyrrolizidine alkaloids. Liver failure, jaundice, and an uncomfortably distended stomach can result after weeks of continuous toxic exposure.[9] In India, unseasonal rains that preceded chronic drought conditions caused moldy food. This was associated with an outbreak of *aflatoxicosis,* a condition that can affect protein metabolism and is a risk factor for liver cancer.[10] In China, a drought caused selenium to enter food, afflicting local communities with hair and nail loss, skin lesions, tooth decay, and nervous system problems.[11] In Ethiopia, consumption of drought-resistant grass pea led to irreversible *neurolathyrism*, a neurodegenerative disease causing lower limb spasticity.[12] In Tanzania and Mozambique, drought conditions were associated with outbreaks of *Konzo*, a devastating neurological disease that causes irreversible paralysis. Water scarcity contributed to inadequately processing cassava, a staple food that must be soaked in water repeatedly for three to four days to safely rid the residual flour of neurotoxic cyanogens. A "short" soaking period of one to two days during the drought was deemed a causal factor for the outbreaks.[13,14]

Insufficient water availability is a defining feature of droughts. Diminished access to water is not the only feature that contributes to the health threat. In addition, the dilution capacity of a body of water is reduced, diminishing the safety of potable water sources. As a water source becomes concentrated, it carries higher levels of chemicals and solid particles per volume of water, lower dissolved oxygen levels, and a higher density of germs that cause infectious diseases. Fewer and smaller water sources during drought results in many more users (humans and animals) for a limited resource, which makes germ transmission and outbreaks of infectious disease more likely. Waterborne diseases that have been linked to drought include amoebiasis, hepatitis A, salmonellosis, schistosomiasis, shigellosis, typhoid and paratyphoid fevers, infection with *E. coli*, cholera, and leptospirosis.[15,16] Skin infections, such as scabies and impetigo, and eye infections, including conjunctivitis, due to a lack of water available for personal hygiene also are correlated with drought.[17]

As soils become increasingly dry during a drought, windblown dust can become a common feature of the air we breathe. During the Dust Bowl of the Great Depression, people suffered from "dust pneumonia," a respiratory condition brought on by inhalation of excessive amounts of dust and dirt.[18] During a drought in Canada in 1987–1989, dust from a dried lake was associated with significant respiratory problems, including coughing and wheezing.[19] Coccidioidomycosis (Valley fever) is a lung infection that is significantly associated with severity of drought.[20]

Vector-borne diseases predominantly carried by mosquitoes, such as malaria, dengue, Zika virus infection, West Nile virus infection, and Chikungunya, can be related to droughts. A scientific study compared mosquito densities in United States wetlands that never dry to densities in wetlands that dry each year. Researchers found that mosquito numbers increased dramatically after large rainfalls that followed natural drought events. The increase in numbers was attributed in part to a loss of insect competitors and decrease in predators (upon mosquitoes) caused by the drought.[21]

Other studies have documented the impact of drought on human mental health. Given the economic stresses that droughts place upon agricultural areas, it's not surprising that anxiety and depression (including disturbed sleep, crying, and fatigue) are common themes, along with feelings of loss and worries about family. One study cited twice the rate of mental health problems in farmers during active drought circumstances compared to farmers not currently in drought.[22] Substance abuse is more prevalent in persons affected by drought.[23]

Drought may force people to relocate. The negative impact of mass movement on wellness can be profound. Researchers studied the health effects of the 1984–1985 drought in Ethiopia, when six hundred thousand affected persons internally migrated from the malaria-free highlands to the humid western region of the country. The settlers were unaccustomed to lowland tropical diseases and therefore had developed little resistance to them; they suffered from higher-than-normal local rates of debilitating parasitic diseases, such as malaria, trypanosomiasis, sand flea infestation, and elephantiasis.[24] When large populations migrate, refugees often are compelled to live together in crowded communities or camps with inadequate sanitation, water, and nutrition. These are ideal conditions for outbreaks of infectious diseases, such as cholera. Migrations also are linked with civil strife, which occurs mostly in populations with poor governance. Drought has been implicated as a threat multiplier in the largest humanitarian crises in recent memory—the conflicts in Syria and Iraq. To worsen the situation, meteorological research has determined that the Eastern Mediterranean Levant region (Cyprus, Israel, Jordan, Lebanon, Palestine, Syria, and Turkey) is likely facing the driest period on record in the past nine hundred years and almost certainly the worst drought in five hundred years.[25] Furthermore, there is some suggestion that human conflict (and bad behavior in general) is more common when the outside temperature is hot.[26]

CATCH A WAVE, AND YOU'RE SITTING ON THE BOTTOM OF THE WORLD

It began as a day like any other at the picturesque unpopulated beach. The "King Tide" surged under a full moon, which lined up with the sun to exert a higher-than-usual gravitational pull. In the distance, cold water lapped at the deck of a remote boardwalk that decades ago had perched a few feet above the water's surface. If the past few years of rising sea level continued apace, then next year, the boardwalk platform would at times be underwater. The restaurants had already abandoned the location, and now the sturdy posts were only occasionally used as a mooring for fishing boats.

"Charlene, be careful. Don't get so close to the water. What if a great white shark jumps up and bites you?" The young mother had an adventurous fourth grader, who looked back and defiantly dipped a toe in the water. She was a feisty little girl. However, like most other small children her age, she was not a particularly strong swimmer or fast runner.

Mom wasn't an earth scientist or oceanographer. Besides, she was focused on her daughter's antics and didn't notice when, in time with the waves striking the surf, the water level under the dock dropped by eight inches, held for thirty seconds, climbed halfway back, and then dropped by a foot. What happened next caught her, and hundreds of thousands of other people living in adjacent coastal communities, unprepared.

A small tidal wave, caused by an offshore earthquake, compounded the local effect of sea level rise caused by melting polar ice. It was already the case in this embattled oceanfront that cliffs had eroded and dwellings collapsed, but these events were skillfully predicted so that people could be moved out of harm's way. But as is almost always the way with a storm surge, a torrential cloudburst onto parched ground that drains into a slot canyon, a rising river in the aftermath of Midwestern summer rains, or a tsunami, extreme sudden flooding catches people by surprise. Overwhelmed by the magnitude of events, the bystanders try to hold their ground. But the water always wins, as relentless tons of liquid engulf defenseless humans.

In this case, the ocean became a monster. Mother and daughter were separated when the small hump in the ocean flowed past them, rolling the surface of the dock like a carpet being flipped for a cleaning. The height of the water they encountered was five feet above normal, a decent wave by surfing standards. They had no way to ride it and were tumbled like rag dolls, driven by the sea onto land, where they struck utility poles and parked cars. Bones were broken. Reflexively inhaling, their lungs filled with sand and debris-laden water. They had only a moment of fear and agony before they became unconscious. If by some miracle they had survived, after four minutes of oxygen deprivation, they would have suffered irreparable severe brain damage and been profoundly disabled for the rest of their lives.

FLOOD

Flood is the most common natural disaster in both developed and developing countries.[27] Floods are expected to increase worldwide in intensity and frequency from both extreme weather events and sea level rise. Their impacts can be deadly and widespread. In the twentieth century alone, dozens of floods have resulted in more than ten thousand deaths apiece.[28] In 2005, Hurricane Katrina struck New Orleans, causing hundreds of deaths and displacing thousands of people. The effects of floods on human health are diverse and vary in time and scope. Direct and instantaneous effects, such as trauma or drowning, occur in the chaos of rising, rushing waters. Days after the flood, the risk of infectious disease spikes due to exposure to stagnant and contaminated water. Months to years later, the toll on mental health from loss and displacement becomes very evident. The nature and extent of impact upon a given population depend on the flood characteristics and baseline vulnerability and resources of the victims, such as infrastructure and isolation, and access to emergency services and post-flood relief resources.

Flooding has been experienced by most civilizations throughout history. It has had such a profound impact upon the human psyche that across all cultures, floods repeatedly have been cited as "proof" of divine disfavor. The most feared outcome is drowning, which is defined as respiratory impairment due to submersion in a liquid. Drowning may not always be fatal, but it is life threatening whenever liquid in the airways hinders respiration, particularly if it immediately results in critical reduction in available oxygen for breathing. Demographic information on flood-related drowning deaths is limited, but we are fairly certain that known risk factors are speed of onset of a flood (e.g., flash floods are more dangerous than are slow-onset floods), water depth (deeper waters portend higher risk), and number of people impacted.[29] Many drownings, particularly in high-income nations, occur when occupied motor vehicles are swept away by floodwaters. The elderly are at great risk of flood-related morbidity due to their physiologic vulnerability, and because they are more likely to ignore recommendations to leave their flooding homes, or even to refuse assisted evacuation.[30] Traumatic injuries can occur from fast-moving and debris-laden floodwaters and include sprains, abrasions, cuts, soft tissue foreign bodies, and bone breaks. These injuries also can occur as people flee harm's way, often with members of their families and/or their possessions in tow. The potential also is significant for injury when displaced victims return to their houses and places of commerce to engage in flood recovery and cleanup operations. Post-flood environments are dangerous. They are full of broken and sharp objects, and hazardous materials. People may unknowingly enter badly damaged residences that are structurally unstable and contain exposed gas pipes and electrical power cables.

In the days after a flood, another peril can emerge. Floods essentially act as a giant environmental blender, mixing industrial wastes, drinking water, and sewage, leaving pools of stagnant and contaminated water in close proximity to humans. The consequence is increased risk for infectious disease. Studies have documented post-flood increases in diarrhea, and specifically in the waterborne diseases cryptosporidiosis, cholera, rotavirus, polio, typhoid, and paratyphoid.[31]

The incidence of mosquito-borne diseases has been documented to increase after floods, particularly in poor, urban settings where dense concentrations of people live near stagnant collections of water (from blocked drains, etc.), which are ideal conditions for mosquito breeding and thus for disease outbreaks. There have been numerous such reports from Africa, Asia, and Latin America.[32] A significant increase in malaria was attributed to the 1982 El Niño event, which caused extensive flooding in several Latin American nations.[33] Another health risk is from leptospirosis, a bacterial waterborne disease that, without treatment, can lead to kidney damage, meningitis (inflammation of the membrane around the brain and spinal cord), liver failure, breathing difficulties, and death.[34]

After floodwaters recede, the emotional trauma remains; for many survivors, life is never the same. The effects on some people's health, relationships, and welfare can be extensive and prolonged, not only from the primary trauma of experiencing a flood, but also because of secondary stressors, such as displacement and the economic pressures associated with rebuilding. As victims attempt to reclaim their lives, property, and relationships, they are more likely than non-victims to experience depression, anxiety, and/or post-traumatic stress disorder (PTSD), and to commit child abuse and suicide.[35] The few studies that examined the effects of flooding on children revealed increases in aggression, bedwetting, and moderate to severe stress symptoms.[36]

WILDFIRE

Hot and dry—they go hand in hand. As temperatures climb and rainfall diminishes, the landscape changes. Grass turns brown, leaves wither and die, and undergrowth is choked off from its life source and becomes brittle. Formerly majestic trees lose limbs, and if the water shortage is profound and prolonged, they die, crack, and fall. Verdant forests become flammable fuel, endless until it burns and is reduced to ashes.

Humans encroach on forests, notably in what is known as the wildland-urban interface.[37] We very often don't build defensible dwellings, even though historical experience lets us know with a reasonable amount of precision the circumstances and risks of a specific location under fire-prone weather conditions. When a dry forest goes up in flames, it spews embers and fireballs, and overruns the terrain and dwellings with ferocity unmatched by any other natural disaster.

NO WAY OUT

"Dry lightning" caused the fire. The storm clouds accumulated a colossal amount of static electricity, which they carried over the hills until the differential between the earth below and sky above provoked a leader stroke, followed by the massive discharge contained within the main channel. In the midst of a forest, the tallest cluster of trees beckoned to the black bottom of a gray anvil overhead, and Thor could not resist. In a single mighty instant, the match struck in the heavens ignited the ground below.

Tom and Sally thought they were prepared. They lived in a log cabin in the woods not far from Bend and had cleared a twenty-foot perimeter around the foundation. Sprinklers were strategically positioned on the roof to soak it down, and the truck was next to the house with a full tank of gas for a quick escape. If fire struck, they would do battle only briefly and leave with their valuables when it became apparent that their dwelling could not be saved. They realized that their evacuation plan was cutting it close, but they knew the forest service roads near their property like the back of their hand.

They had every chance to get out. The fire started five miles away, and the smoke came early, pushed by the wind. It was dense and turned the incandescent sun into an orange ball that you could stare at without having to look away. A neighbor drove up to their home with a truck full of his belongings and told them that everyone in the area had been given instructions to evacuate—no point taking chances. No rain was in the forecast, and the winds would shift many times. Helicopters carrying huge bags of water pulled from lakes and planes spraying brightly colored fire retardant chemicals flew overhead. There was no mistaking the gravity of the situation. War was being waged on this fire.

Even though it's the right thing to do, nobody wants to abandon his home. Tom had been a volunteer firefighter and should have known better. Instead of exercising caution, he mistakenly thought they could tough it out for another fifteen minutes. "We'll be right behind you," he assured his friends. That would be the last time they saw each other.

The wall of fire came hard, fast, and unexpectedly from two directions. The inferno burned or melted everything in its path. By the time Tom realized that Sally was not responding to his shouts, she was face down in the backyard, crushed by a fallen tree limb. In less than five minutes, she was burned beyond recognition. In a desperate attempt to save himself, Tom quickly dug a shallow pit in the ground, knelt down, and covered up with an old silver "space blanket." In a lesser fire, this might have saved his life, but this was no lesser fire. The scalding air and flames made quick work of Tom, and destroyed his home and thousands of acres of surrounding forest. It was one of dozens of fires triggered by the same weather system, in a pattern that would be repeated more often as weather and fuel combined to create a perpetual fire hazard.

As climate change redefines the new "normal" in our weather patterns, wet places may become wetter and dry places drier. Warm, drier weather is a driver for seasonal wildfires; forests and soil hold less moisture, and there are less water and snowpack to act as a buffer against fire. In the western United States, we've seen a pattern that is now replicated worldwide. Wildfire numbers are increasing, and fire seasons are growing longer.[38]

By far the greatest number of people is put at risk for experiencing one or more effects of wildfires by large swaths of enormous air masses degraded by smoke. These can be the size of an entire western American state during peak fire season and can travel thousands of miles.[39] Wildfire smoke is a mixture of gas and fine particles from burning vegetation and/or building materials and can contain large amounts of toxic carbon monoxide, nitrogen oxides, ozone precursors, particulate matter, and even poisonous mercury precursors.[40,41] Smoke exposure can cause coughing, shortness of breath, sore throat, eye irritation, runny nose, and sinus pain. It can also trigger wheezing. For persons with reactive airway disease, smoke exposure can precipitate a full-blown asthma attack. Wildfires can burn for months. Persistent smoke generation can place downwind inhabitants at risk for chronic lung irritation, headaches, and fatigue. Such a situation has been linked directly to increased emergency department visits.[42,43] Persons with preexisting heart or lung conditions risk further stress and the necessity for medical interventions and/or hospitalizations. The very young, elderly, and pregnant women are most at risk from wildfire smoke, because their lungs and the developing fetus, respectively, are the most sensitive to the injurious effects of smoke.

Extreme weather events are disruptive and almost always hazardous to human health. From the perspective of those who need to decide whether to take actions that will diminish the likelihood of such events, it is mandatory to understand what they are, how many people they affect, and what would be the benefit to humankind if we were able to cause them to be fewer, less intense, and of shorter duration. Knowing how to survive a wildfire or flood is not nearly as good as not having to endure them.

· 5 ·

Vector-Borne Diseases

MUCH MORE THAN A NUISANCE

Long Beach Island had been a haven for Mark every summer since childhood. His father meticulously packed the car trunk with coolers of food, beach chairs, kites, and everything else the family would need for two weeks' vacation at the beach. Surf City was a congregation of young and old who loved the sand dunes and salty air, smell of the ocean and Coppertone, and breeze from the bay. When he was a small boy, Mark would bait his crab trap with chicken legs and lower it off the dock, hoping to lure a full house of blue crabs. When he became a teen, the social scene was much more important than fishing for flounder in front of Barnegat Lighthouse. He met his life mate when he helped her navigate a rip tide safely to shore. Years later, when they had children, the Jersey shore was his sentimental favorite. It warmed his heart to watch his small children build sand castles and try to protect their imagined fortresses with collapsing moats from the foam in front of the incoming waves.

One bad aspect of summer vacations that had worsened was the frequency and ferocity of biting insects. The enormous green-headed flies that appeared with an incoming "land breeze" left angry, red swellings. Tick infestations were sporadic, but mosquitoes had become omnipresent. Despite municipal spraying with insecticides, the mosquitoes bred in pools of brackish water left behind by thunderstorms, lurked in the sandy weeds, and infiltrated houses through holes in screens that covered windows and doors. Outside or inside, it didn't matter. If you summered at the beach, you were bitten.

Mark had returned home after his beach trip. Three weeks later, the fever first struck. He had a headache, began to sweat, developed aching muscles, was enormously tired, began to vomit, and suffered from diarrhea. A "cold" was going around the office, and he assumed he had it. But this wasn't a cold. The headaches became more intense and lasted longer, he began to have shaking chills, his fever soared, and he would remain sick for hours at a time, with a few days of remission between episodes of illness.

When his fifth bout of total debilitation put him in bed for a day and night, he sought medical care. By that point he had constant abdominal pain, was having trouble breathing, and noticed blood in his urine. He was fortunate to have an astute doctor who had become accustomed to diagnosing diseases that had formerly been considered "tropical." He promptly ordered a blood sample to be taken from Mark, and personally examined it under the microscope. He saw the parasites of malaria. To confirm what he suspected, he used a rapid diagnostic test. His diagnosis was correct. From that point forward, Mark was treated skillfully, which was critically important because he had falciparum malaria, statistically the most lethal variety. He required hospitalization, powerful antimalarial drugs, and intricate intravenous fluid management. Worldwide, the incidence of malaria followed the mosquitoes, and what formerly had been a disease suspected in travelers returning from South America, Africa, or Asia was becoming commonplace in northern locales.

OVERVIEW

A warmer world will create newly hospitable habitats for tropical and subtropical insect vectors and the diseases they carry. Historically, disease-free areas have been protected from becoming hazardous by cold environmental temperatures. That is, with extreme low temperatures of winter, insect (and in particular, mosquito) populations are decimated. However, as the average global temperature creeps up, mosquitoes will thrive longer and reproduce more successfully at higher latitudes and altitudes. In the Northern Hemisphere, they are spreading northward and increasing their natural habitat. Vector-borne diseases (diseases carried by other species, such as mosquitoes, ticks, and fleas) take an enormous toll on human health. Malaria, dengue, West Nile virus disease, Lyme disease, and Zika virus infection now and increasingly will account for much of the global burden of disease. The World Health Organization estimates that every year more than one billion cases of vector-borne disease result in more than one million deaths.[1]

The vectors that carry these lethal diseases are classified in the animal phylum *Arthopoda*. These vectors are hardy and adaptable creatures, and global warming promotes their increase in numbers and geographic range. Increasing temperature increases the rate at which the infectious organisms the vector carries can replicate within the host (e.g., mosquito), which directly correlates to infectivity to humans. Mosquitoes thrive in warmer temperatures, which generally boost their rates of reproduction and the number of blood meals they take, extend their breeding season, and curtail the maturation period for the infectious organisms they disperse.[2] In a warmer climate, mosquitoes breed and bite more, and so spread more disease.[3]

MALARIA

Malaria is caused by protozoan parasites of the genus *Plasmodium* that are transmitted by *Anopheles* mosquitoes. These mosquitoes have an intricate life cycle that involves both the mosquito and human host. After a person has been bitten and infected by a malaria-carrying mosquito, an incubation period of one to three weeks follows before disease symptoms begin (which is why many travelers show symptoms only after returning home). Upon entering the human bloodstream after a mosquito bite, the parasites travel to the liver, where they multiply. From there, they inhabit and destroy red blood cells, which coincides with the onset of severe symptoms. High fevers, muscle aches, shivering, nausea, and vomiting are common. Young children are at risk for seizures. Symptoms can involve multiple organs; and infection leads to severe anemia, kidney failure, fluid in the lungs, low blood sugar, altered mental status, and coma. Symptoms may reappear spontaneously after months of good health. Malaria can be cured, but access to the best (or sometimes any) medicine remains a problem worldwide, as does drug resistance. In certain parts of the world, most of the historically successful antimalarial treatments are no longer effective because of this resistance, limiting the treatment options in endemic areas.

Malaria is currently responsible for 220 million cases and 660,000 deaths worldwide annually.[4] It represents a significant percentage of the global burden of disease and is skewed toward the most impoverished populations. Increased global mean temperature is expected to expand mosquito habitat, accelerating disease transmission. Numerous studies (e.g., close observation of areas affected by the El Niño phenomenon) have correlated the link between malaria outbreaks and temperature, humidity, and climate fluctuations.[5,6,7] A good deal of data that support our understanding about how a changing climate can affect malaria comes from research performed in the east Africa highlands.[8] The cooler highlands of this equatorial region historically have had a tempering effect on malaria transmission, because mosquitoes cannot thrive as effectively there as within the adjacent tropical lowlands. However, since the 1980s, malaria cases in the highlands have increased in number.[9] Because of the previous low incidence of disease, the highland human population had a low resistance to malaria, so when the disease entered this habitat, susceptibility with a propensity for a more severe disease course was relatively high. In western Kenya, malaria epidemics have increased their range from three to fifteen districts, with a simultaneous disease incidence increase from 20 to 60 percent in that region.[10]

The relationship between humans, mosquitoes, and the malaria parasite determines disease transmission. Health-care providers seek to disrupt this relationship in any way possible. A warming climate will act to facilitate it. Increased environmental temperature will bolster development and reproduction of mosquitoes, and allow them to thrive longer. This will enable them to have more time to ingest human blood meals, thereby increasing the disease vector capacity.

When predicting the future, the human element is less certain. Many factors are used to determine malaria disease risk, including land use, population growth and urbanization, human migration, economic development, healthcare infrastructure, and governance (e.g., financing, vaccine availability, using insecticides). However, as global warming occurs, we can look at the excellent data on the correlation between seasonal weather fluctuations and malaria outbreaks to refine our predictions. Unfortunately for us, a warmer world generally favors mosquitoes and, therefore, the diseases they carry.[11]

ZIKA VIRUS DISEASE

A CRUEL BITE

Bugs were a fact of life on the Outer Banks. The mosquitoes came off the dune grasses. They landed in stealth, then bit suddenly and ferociously, drawing blood and creating a welt that itched for days. The itching was intense, and if you happened to be a person who had been sensitized to mosquito saliva, the bumps became mountains you scratched for a week. However, you didn't get sick. On the East Coast of America, you didn't get dengue, Chikungunya virus disease, or anything like that. Those things happened in other countries. Our mosquitoes were a nuisance, but they couldn't kill you.

Frank and Nancy had been vacationing in North Carolina for as long as they could remember. Their children and grandchildren loved it. Days on the beach, running in the sand, croakers and hushpuppies for dinner, and then a big ice cream cone. It was terrific. Of course, the flies and mosquitoes were still there, but no more spraying. It was up to individuals to protect themselves with insect repellents. The advice was out there—stay out of the sun and wear your sunscreen—cover your skin with repellent and watch out for the bugs.

Their grandson Phil married his sweetheart in June and decided to visit his grandparents in August. It would be a wonderful time—childhood memories galore. They knew all about skin cancer, so they dosed up with sunscreen, but in a rush to get to the beach, they went light on the repellent. After they played in the surf for an hour, they were unprotected. That evening at dusk, they walked to enjoy the moonlight. They wore shorts and T-shirts, and had the living daylights bitten out of them by the mosquitoes.

When Frank and Nancy found out they would be great grandparents, everyone rejoiced. But as had many expectant parents before them, they felt an underlying unease. Their fears were confirmed by an ultrasound in the second trimester of the pregnancy. The Zika virus claimed another victim.

Zika virus disease is a vector-borne disease, which like malaria is transmitted by mosquitoes. Although the disease was first identified as far back as 1947, it reached epidemic levels in 2016 as it blossomed into a full-blown public health crisis. In just eight months, the number of cases ballooned to just under four million patients affected in more than twenty-three nations.[12] Although there is much debate about what prompted emergence of the epidemic, consensus is that it may well have been augmented by heavy rains and heat.[13] The year 2016 was an El Niño year, which meant above normal temperatures and precipitation in Central and South America, where the current Zika epidemic first emerged. The bottom line is this: mosquitoes carry the virus, and as they move north, so will Zika virus infections.

Zika virus disease is frightening because it has profound adverse consequences for health and has spread so fast. It's estimated that for every one million people at risk, a staggering 70 percent may become infected.[14] One in five infected persons will suffer symptoms of mild viral illness (fevers, chills, muscle aches, and fatigue).[15] The horror resides in the number of cases of devastating microcephaly (abnormally small head with severe developmental delay) births and the neurologic disorder known as Guillain-Barré syndrome (GBS), a disorder in which one's immune system attacks parts of the nervous system, causing severe weakness and ascending paralysis. For every one million people at risk, the rates for microcephaly and GBS are 2,625 and 140 individuals, respectively.[16] Considering that the virus is now endemic throughout large portions of the Western Hemisphere, 2.2 billion people currently are at risk.[17] This number very likely will increase until a viable preventive vaccine is introduced.

The major source of Zika virus transmission is via mosquitoes of the genus *Aedes*, the same genus that transmits dengue and Chikungunya viruses. These mosquitoes range throughout most of the subtropical and tropical world, including the southern United States as far north as Washington, D.C. Like the *Anopheles* mosquitoes that carry malaria, these mosquitoes lay eggs in and near standing water. Mosquitoes may become infected when they feed on a person already infected with the virus and then spread the virus to other people through bites. The vast majority of people become infected through mosquitoes, but case reports also have shown direct transmission through sexual contact, mother-to-fetus transfer, and blood transfusion.

Until recently, the Zika virus has been relatively unstudied, so compared to other viruses and diseases, much less is known about how climate change might influence its future spread. We are safe to assume that research on similar viruses can inform us about what climate change might mean for the mosquitoes that carry Zika virus. In this regard, dengue, described next, is an excellent surrogate for epidemiological purposes.

DENGUE

HARMFUL HEADACHE

"Not again," Roger thought. This was Haiti all over again, but how could that be? He was in West Virginia. Years earlier, he had responded with his church group to the poorest Caribbean nation to assist in building shelters for displaced persons. He had been warned to wear insect repellent, cover as much exposed skin with clothing as possible, and sleep under a bed net. Malaria was the insect-transmitted disease they feared the most, and so they took anti-malaria drugs for prophylaxis. That worked. Very few of them had been stricken with malaria. But no drug could prevent dengue. From a pool of stagnant water he had been bitten by an *Aedes* mosquito, which inoculated him with the virus. He suffered a mild viral illness and thought he was immune for life. However, that was not the case. There are four different *serotypes* (or variations) of dengue virus (DEN-1, DEN-2, DEN-3, and DEN-4), and there appears to be little cross-immunity, so a person may be stricken with dengue four times in his life, with each subsequent infection generally being worse than the preceding ones.

Back in the United States, volunteer work took him to every corner of the country. Helping the less fortunate gave him joy and a sense of accomplishment. This time, it infected him with a disease. The mosquito larvae flourished in the rainwater that lingered in the discarded truck tires that ringed the perimeter of the garbage dump. They matured into flying creatures. When they were sufficiently mature to seek a blood meal, they swarmed to the carbon dioxide emanating from the skin of unsuspecting volunteer construction workers who had come to offer their services to the poor people in need of shelter.

The incubation period following the mosquito bite was five days. Without warning, Roger suffered a splitting headache that emanated from behind his eyes, a sore throat, fatigue, coughing, and a high fever. He thought he might have the flu, because of the chills, muscle aches, and nausea and vomiting. But it was worse than that. His bones seemed to hurt, and then he noticed a fine, red, itchy skin rash on his upper arms, legs, and trunk. At first the rash spared his face, palms, and soles, but then spread to these as well while becoming darker and more confluent. After a few days the rash began to fade, but he suffered cycles of a few days of fever, then one to three days without fever, then fever again.

He wasn't alone. A coworker began his illness with similar symptoms, then further worsened, with emotional depression and then a seizure. The teenage daughter of the camp pastor collapsed with severe abdominal pain and persistent vomiting. Her dehydration was compounded by bleeding problems, including nosebleed, bleeding gums, and bloody vomiting, all of which led to profound weakness. The virus attacked her vital organs, so she had breathing difficulties and failure to maintain her circulation, which culminated in extremely low blood pressure. In just a few days after her first fever, the unfortunate young woman developed a diffuse, dark purple, blotchy rash caused by bleeding into the skin that indicated a life-threatening situation.

Dengue is named colloquially (and appropriately) "breakbone fever." Like other mosquito-borne diseases, dengue is projected to increase in range from climate change. Modeling studies suggest that the projected global warming of 2°C (3.6°F) by the year 2100 will result in a net increase in the potential latitudinal and altitudinal range of dengue and increased duration of the transmission season in temperate locations.[18] Dengue predominantly is an urban disease, and therefore more than half of the world's population, mostly in tropical areas, lives in areas of risk.[19] Dengue transmission occurs year round but has seasonal peaks during months with high rainfall and humidity.

Two clinical manifestations of the disease can be life threatening: dengue hemorrhagic fever (DHF) and dengue shock syndrome (DSS). DHF warning signs include severe abdominal pain, persistent vomiting, marked change in body temperature (either fever or an abnormally low temperature, i.e., hypothermia), altered mentation, and/or bleeding manifestations. This abnormal bleeding is a result of platelet dysfunction, which interferes with normal blood clotting. The most common hemorrhagic manifestations seen with dengue are mild and include easy or spontaneous bruising or tiny skin hemorrhages, nosebleeds, bleeding from the gums, and blood (usually invisible to the naked eye) in the urine. A positive "tourniquet test" is now part of the World Health Organization criteria for diagnosis of dengue. This is a test for capillary fragility. To perform the test, the clinician inflates a blood pressure cuff on a patient's arm for at least five minutes. If there are many obvious clustered tiny bleeding spots below the elbow on the skin of the lower arm after the cuff is deflated, the test is positive. The more concerning bleeding manifestations of dengue include vaginal bleeding, vomiting blood, tarry black (blood-laden from gastrointestinal bleeding) stools, and intracranial (within and around the brain) bleeding, any of which can become life threatening.

Dengue shock syndrome (DSS) is the other feared clinical entity this virus causes. It is defined as any case that meets criteria for DHF with features of shock, which is inadequate supply of the blood (and the energy and oxygen that it carries) to the vital organs. Patients with DSS have a rapid weak pulse and low blood pressure, are restless, and often have clammy skin with cool limbs. Fatality rates of patients with DSS can be 10 percent or higher, but fortunately, with early recognition and intensive care, can be less than 1 percent.[20]

WEST NILE VIRUS

West Nile virus (WNV) disease is another mosquito-borne virus that will be affected by a changing environment, as its range is also predicted to spread. About 80 percent of persons infected with WNV remain asymptomatic.[21] Those who become symptomatic have a typical viral syndrome of high fever, chills, headache, muscle aches, and rash. In less than 1 percent of cases, more severe neurologic debilitation is manifested as altered mental status, meningitis, tremors, convulsions, vision loss, numbness, and paralysis.[22] These terrifying symptoms can last weeks or months, or even become permanent. The elderly are particularly at risk for neurologic complications from WNV, as are persons with specific medical conditions, such as cancer, diabetes, hypertension, kidney disease, and organ transplantation. Approximately 10 percent of people die who develop neurologic infection due to West Nile virus.[23]

WNV was first diagnosed in the West Nile subregion of Uganda in the late 1930s, but it was not seen in the United States until 1999. Since then, more than thirty thousand cases have been reported in the United States.[24] Human infections usually occur in the late summer and early autumn after sufficient virus accumulation in mosquitoes that have acquired the virus from species of birds that act as "amplifying hosts." These birds are carriers in which the level of virus can become so high that a mosquito that feeds on them becomes infectious to humans.

The mosquito genus *Culex* prefers to feed on birds, but it bites humans in an opportunistic fashion. As the number of bird hosts declines in the late summer and early autumn (due to seasonal migration), *Culex* feeding preference shifts to mammals, increasing the risk of infection to humans. However, as with other mosquito vector-borne diseases, the strongest environmental predictors of WNV disease risk are temperature and precipitation. Data from North America, Russia, and Israel all have correlated higher temperatures with disease outbreaks several weeks later.[25,26]

Although heavy spring rainfall may boost mosquito populations later in the summer and increase disease transmission, heavy rains can likewise flush mosquito larvae from breeding locations and decrease their numbers. Although WNV mosquito populations thrive in subtropical conditions, extreme heat has a "heat threshold" that ultimately decreases their populations.[27] What can be ascertained from these data is that WNV transmission and its impact upon human illness through infection are strongly linked to climate variability. In this regard we can expect an increase of WNV prevalence at higher latitudes, where the virus will encounter both mosquito vectors and a human population that historically has not been exposed to the disease, rendering it highly susceptible to outbreaks. This susceptibility is due both to behavioral factors in communities that previously have not had to plan for the presence of this disease, and the fact that mosquitoes will be biting mostly people naive (i.e., without immunity from a prior infection) to WNV disease.

LYME DISEASE

Lyme disease is caused by an infectious organism carried by ticks. In the United States and Europe, the organism is the bacterial spirochete *Borrelia burgdorferi*. The natural reservoirs of the bacteria are small rodents and birds. Ticks of the genus *Ixodes* act as disease vectors to pass spirochetes from small mammal reservoirs to infect humans. Empowering the disease further are large mammals (such as deer), which play a key role in supporting *Ixodes* tick populations.

Lyme disease has many stages. Its onset usually is insidious with vague symptoms, such as fever, headache, and fatigue. Often, but not always, a characteristic "bulls-eye" rash known as *erythema chronicum migrans* appears. If Lyme disease is recognized and treated promptly with antibiotics, symptoms dissipate without further consequences. However, if treatment is improper or absent, spirochetes eventually proliferate in the body and spread to joints, causing debilitating arthritis, as well as to the nervous and cardiac systems, where symptoms can be as varied as paralysis on one side of the face (Bell's palsy); or heart block, which is disruption of electrical transmission within the heart, leading to abnormal heart rhythm, and that might need to be treated with an implanted pacemaker. Other symptoms of Lyme disease include chronic and severe headaches, depression and mood changes, memory loss, dizziness, shortness of breath, and neuralgias, which are nerve-generated jolting pains that sometimes precede paralysis. Chronic neurologic symptoms occur in up to 5 percent of untreated patients.[28]

More than thirty thousand cases of Lyme disease are reported each year in the United States, making it the most frequently reported vector-borne disease and in the top ten most common infectious diseases in the country. More than 95 percent of these cases occur in fifteen states in the Northeast and upper Midwest.[29]

Models that incorporate climate change with other environmental factors (e.g., suburban growth, deer populations, reforestation, and water usage) affecting tick distribution show the geographical range of the tick vectors and the areas at risk for Lyme disease likely to expand significantly in the coming decades.[30] Much of this has been due to northward expansion into southern Canada, an area that previously has been too cold to support *Ixodes* populations.[31] Lyme disease is also expanding into the American South and mid–Atlantic at the same time, with large increases in Delaware, Maryland, and Virginia.[32] This reinforces a multifactorial etiology for the spread of the disease beyond temperature change, which makes sense given that the disease is maintained in a complex cycle involving ticks, multiple bird and mammalian species, and humans. Undoubtedly, sustained increases in average temperatures will lead to more infectious disease spread in higher latitudes and elevations, where human populations have little to no historical interaction with these pathogens and thus little immunity.[33]

The main issue with vector-borne diseases and climate change is the geographical range in which the insect vectors will flourish and have access to their animal hosts and human victims. Climate change perhaps will not create new diseases, but it will broadly increase the human populations at risk, and thereby make the imperative for effective vaccines and protective guidance more urgent. In addition, given the increasing amount of global travel and, indeed, movement of entire populations while in close proximity to one another, we can expect to see more rapid spread of disease within groups and failure to contain rapid spread of infectious diseases.

· 6 ·

Mental Health

$\mathcal{E}$ffects on mental health related to global climate change will arise from multiple causes, both directly and indirectly. Direct effects include distress immediately following traumatic events brought on by severe weather episodes: tsunamis, hurricanes, wildfires, and other climate-related environmental disasters. Emotional distress also may be created by desecration of land and crops due to weather conditions, such as flooding and droughts that force population displacement when local conditions become untenable. Indirect effects include anxiety and depression related to physical maladies caused by climate change, or from the negative effects that climate change can create, such as degradation of the social, economic, or emotional fabric of a community.

Extreme weather events, such as tsunamis (e.g., Indian Ocean, 2004), hurricanes (e.g., New Orleans, 2005), and long periods of drought (e.g., Syria, 2006 on) have and will continue to cause permanent mass displacement of large populations due to destruction of homes and disrupted communities.[1] Property damage can be a profound blow to a person's psyche, challenging the "sense of self," thereby causing bereavement over loss of personal investment, and also endangering intimate relationships and cherished memories.[2]

During the migrations that follow such disasters, families risk disruption. Social connections are easily broken. Migration itself is a health stressor; the possibility of separating family members, and physical injuries or fatalities of loved ones is further cause for mental anguish.[3] According to the Intergovernmental Panel on Climate Change, by the year 2050, 150 million refugees may be displaced by coastal flooding, shoreline erosion, and agricultural degradation.[4]

Because displacement often becomes a permanent condition, groups may have a difficult time reestablishing livelihoods and communities. The struggle

47

to adapt to new environments under such traumatic circumstances is a situation ripe for chronic mental health disorders, such as depression, anxiety, and suicidal ideation, to flourish. Having to coexist with new, unfamiliar, and sometimes hostile neighbors adds to this stress, even more so with the tensions between native populations and recently emigrated groups, and exacerbated by differences in race, religion, language, and culture. Such cultural friction has been shown to result in low self-esteem, anxiety, and depression.[5]

After a severe weather event, the mental health impact can linger for years. The prolonged effects can manifest in myriad ways: post-traumatic stress disorder (PTSD), anxiety, depression, acute psychosis, suicidal thoughts, and substance abuse. Following Hurricane Katrina in 2005, a "high prevalence of psychiatric morbidity" occurred in the affected population.[6] A survey of 1,043 affected residents taken thirty days after the hurricane revealed that 49.1 percent of the group was experiencing anxiety-related mood disorders.[7] In a longer-term evaluation of the mental health impact caused by Katrina, 815 pre-hurricane residents were sampled, and the prevalence of PTSD, as well as attempts at self-harm and suicide, had increased over time.[8]

For persons who experience a traumatic weather event and previously have been diagnosed with a mental disorder, the preexisting condition may flare and become worsened, and mental disorders that had been considered mild risk progression into severe ones.[9] To put this into perspective, across the world population, the presence of mild and moderate common mental disorders (anxiety, depression, PTSD) is about 10 percent.[10] When accounting for the number of people who will be exposed to extreme weather events worldwide, it is probable that this will rise to 20 percent, which is a tremendous number of people burdened with mental illness.[11] Adding insult to injury is the unfortunate perception that mental health illness is something to be ashamed of, and should not be discussed to avoid embarrassment. This misconception prevents the sick from seeking treatment, minimizes available community health resources for mental health, and keeps a burden of disease unnecessarily entrenched within a society.

A SINKING FEELING . . .

Climate change invokes many difficult emotions among affected people and nations. For inhabitants of physically low-lying island nations, they have a sinking feeling, because they are literally sinking. They are now coming to terms that their progeny will inherit a new type of refugee crisis; they will be fleeing from becoming submerged under a rising ocean. Fear and apprehension pervade their national consciousness, which is an understandable sentiment when the world is telling them that their countries may soon be underwater.[12] Kiribati, a nation of thirty-three coral islands in the central Pacific spread out over an area larger than India, is among the most vulnerable. Kirabati's government assessment is planning for the eventuality that the Tarawa atoll, the nation's capital, will become uninhabitable within a generation.[13] Tuvalu is another Pacific island nation at risk. With a population of more than ten thousand people, its highest point is only fifteen feet above sea level, and during tidal surges, all of Tuvalu temporarily is submerged.[14,15] The highest point in the Maldives, a country made up of coral atolls in the Indian Ocean, is less than eight feet above sea level. Current predictions are that Tuvalu will disappear in the next fifty years, and the Maldives in the next thirty years.[16] Neighboring islands will follow, affecting up to 9.2 million people throughout the Pacific Ocean's twenty-two island states and 345,000 people in the Maldives.[17]

Inevitably, the saga of these "climate refugees" will play out like the mass human displacements we see today because of violent conflicts or sudden natural disasters. The consequences of the permanent disappearance of a nation's entire being is a tragic fulfillment of our environmental crisis. People may have existential emotional crises. Throughout the Pacific region, "population transfers" already are taking place to resettlement communities in New Zealand and Australia. The United States has granted citizens of the Marshall Islands the right to live and work visa-free on its shores, and Fiji has pledged to accept climate refugees from Kiribati should the need arise.[18]

Although people can change their addresses, nations cannot yet do this. The physical and mental health challenges of cultural assimilation on such a massive scale will be formidable. How will a fifth-generation fisherman from the Maldives fare in a new community in the Australian outback? How will a Tuvaluan woodworker adapt to rural New Zealand? The mental health stressors to even the most resilient of these people will be demanding and likely will manifest as depression, anxiety, PTSD, and even suicidal thoughts.

LINGERING EFFECTS OF CLIMATE CHANGE ON MENTAL HEALTH

Physical illnesses from climate change cause negative health consequences. These physical maladies likely will be accompanied by mental disorders if quality of life is hindered through illness of self or a loved one. Profound disruptions to work, home life, or relationships that formerly were part of one's mental health equilibrium before a traumatic event can, at the very least, bring on common forms of psychological distress. People with chronic illnesses have a substantially higher risk for mood and anxiety disorders and, sometimes, suicidality.[19] The timing for this state of affairs is not good, because the WHO's most recent Global Burden of Disease report ranked depression among the top three most prevalent causes of worldwide disability.[20,21] Competition for resources and land occupancy may become more entrenched in our emotional landscape as global warming continues to desiccate arable land and clean water. If resources become hard to attain, then the risk of civil strife and conflict will rise, and this will become a driver of mental illness.

In places of profound chaos, many refugees simply flee as a means to keep themselves and their families safe. As seen by the mass exodus to Europe over the past several years, these journeys pose significant peril. People migrating on short notice are frequently ill equipped and unprepared for the strenuous journey. High fatality rates are often part of unplanned mass migrations.[22] Whether fleeing bad politics or bad weather, unplanned and inadequately resourced migration is fraught with physical and emotional health perils.

LUCK RUNS OUT

Farooq was alive. That made him one of the lucky ones, yet he felt like the unluckiest person on Earth. He was raised on an olive farm in Syria and had a happy childhood, surrounded by a loving family with dozens of relatives in a prosperous village. All of that began to change ten years ago, when his country descended into what eventually seemed like hell. It began with a drought. The desiccated farmlands promptly failed. Intense competition for irrigation deprived his family of the water needed for their cotton crop. Yields declined until Farooq and his cousins realized they could no longer sustain themselves in their homeland. As did innumerable others in the same situation, they decided to leave.

Pundits called it the Arab Spring. For those struggling to find food and shelter, it was not a path to prosperity. It was a daily fight to survive. The fortunate ones made it to Bavaria. Others were not so lucky. His grandmother, unable to walk another step, sat down on a roadside boulder in Macedonia, quietly closed her eyes, and never got up. A young niece died in her sister's arms as they both drowned when their overloaded boat capsized attempting to reach Greece.

Two years later and likely forever, the demons will remain. Asleep and drenched with the sweat of nightmares, Farooq relives his journey and the inhumane deprivations. He thrashes and strikes his wife, wide awake next to him. In the daytime, unable to quiet his mind, he immerses himself in inane tasks because he has not regained the attention span to learn a new profession. Once a successful farmer, he relegates himself to rebuilding automobile engines in the silence and seclusion of early morning, simply to avoid contact with people. Physically safe for the time being, he is nonetheless hypervigilant for any crisis, so is continuously traumatized. Despite his relatively decent new surroundings, he fails to embrace a new life for his family, because he is so emotionally impaired.

CULTURAL ANXIETY

We conclude this particular discussion with a mental health stressor that is pervasive and insidious. It is a sense of cultural anxiety that is worsened by discussions about climate change. Whether they come from nightly news reporters or politicians, we seem to lack consensus about how best to determine the truth and then act accordingly. The discourse is not always civil, which is unfortunate, because other than differences of opinion, there is no fundamental reason for blocking communication and collaboration. We need to encourage evidence-based science, appropriate investigations, bona fide analyses, and a collaborative approach to discourse related to environmental issues and global climate change, with the emphasis on "global." This approach will be far more productive than adopting a confrontational approach or one of helplessness, fear, and a sense that we are doomed. We should determine if and where we can intervene for improvements. We must assume that we are not past the point of no return. That line of reasoning will protect our mental wellness and support the emotional energy we need to approach a situation that calls for cooperation, collegiality, and foresight.

Part II

CLEAR AND
PRESENT PATHOGENS

· 7 ·

Air Degradation

$\mathcal{H}$umans maintain their most fundamental and ceaseless interaction with the environment through the air they breathe. The atmosphere that envelops Earth holds an essential element for human survival: oxygen. The average person inhales approximately twenty thousand breaths per day, thereby exchanging roughly ten thousand liters of air through the lungs. This is an unavoidable daily exposure to the ambient environment. Even in our most sedentary moments of rest or sleep, our metabolic activities fueled by the energy required to sustain life necessitate a constant supply of adequate oxygen.

Our lungs provide the means to extract precious oxygen from air. However, the exchange of oxygen-rich gas (inhaled O_2, or oxygen) and gaseous carbon dioxide waste (exhaled) is a far more intricate process than the simple flow of air. The lungs are composed of a tissue matrix that exposes air in the bronchial tree to a dense network of microscopic blood vessels. These tissues ensure gas exchange over a surface area of approximately 160 square meters in a single adult person, an area comparable to that of a tennis court. Air contains predominantly oxygen (21 percent), nitrogen (78 percent), and carbon dioxide (less than 1 percent). Certain inert gasses, such as helium, neon, and argon, comprise the remaining natural components of air. Unfortunately, because of human industrial activities, air also contains a large number of man-made pollutants, such as carbon monoxide, various nitrogen oxides, and sulfur dioxides. Given the capacity of the human body to absorb gaseous chemicals, particulates, and microorganisms through the act of breathing, air quality is intricately linked with human health.

Despite recent successes to improve air quality throughout the world, climate change is pushing back against these gains. This combines with rapid industrialization in developing nations potentially to outpace any progress made to curtail air pollution. More than half of the United States' human population lives in counties that do not meet national ambient air quality standards as set by the Environmental Protection Agency.[1] The air quality in many other countries is grimmer.[2] In China, air pollution is felt to be a major threat to public health, and scientists have called for urgent action to reduce pollution and improve air quality.[3] In the United Kingdom, it is estimated that air pollution arising from burning fossil fuels accounts for forty thousand premature deaths annually.[4] Earth's atmosphere knows no national boundaries, so the threat of airborne toxins is logically and constantly global. We are in a worldwide struggle to clean up what has become a universally held concern.

The two significant pollutants that are most susceptible to a warming climate are *particulate matter* and *ground ozone*. Climate change degrades air quality in other ways as well, such as the contributions of smoke from wildfires, dust from droughts, and allergenic mold from flooding. All of these risks coincide with a demographic shift in the susceptibility of global populations. This will render many more persons vulnerable to air pollution exposures. With rising rates of asthma, diabetes, and obesity, as well as an increasing number of elders, we will have a world with more persons (on a percentage basis) who have less physical reserves and innate physiological defenses to withstand the impact of reduced air quality.[5]

PARTICULATE MATTER

The catchall name for particles suspended in the air is *particulate matter* (PM). The inhaled foreign substances cause not only diminished lung function but also long-term effects of scarring and cancer. Our respiratory tract cannot compensate for many types of prolonged exposure to airborne toxins that cause declines in pulmonary function. Particulate matter has been linked to heart and lung diseases, including heart attack, asthma, bronchitis, and chronic obstructive lung disease. One study has also linked PM exposure to accelerated atherosclerosis ("hardening" of the arteries), a risk factor for coronary artery heart disease.[6] Another sixteen-year study qualified the risk further, concluding that fine particulate and sulfur oxide–related pollution was associated with increases in all-cause, lung cancer, and cardiopulmonary death rates (4, 6, and 8 percent increase, respectively).[7]

The origins of PM stem from many human-devised processes. *Primary particles* originate directly from sources such as smokestacks, fires, and construction processes. *Secondary particles* result from airborne reactions of chemicals, such as sulfur dioxide and nitrogen dioxides, emitted from industrial plants and automobiles. Given the variety of origins, the makeup of PM likewise is diverse. Sulfates, nitrates, ammonia, sodium chloride (salt), carbon, dust, and water are present as various mixtures of solid and liquid substances suspended in air. Climate change is expected to contribute significantly to PM through elevated temperatures, wildfires, and airborne pollen. However, future models of PM air burden are uncertain, given increases in industrialization and international commitments in emissions reductions. Despite this uncertainty, the significant existing health impact of PM causes it to be cited consistently as a major potential health risk from climate change.[8]

Particulate matter can be categorized by size, origin, chemical composition, atmospheric behavior, and method of measurement. A convention by which we understand PM is the size of the particles:

- **PM$_{10}$** *particles* are less than or equal to 10 microns in diameter. Particles larger than this size generally do not reach the lungs because they are "captured" by the upper respiratory tract (nose, larynx, and upper bronchi).
- **PM$_{10-2.5}$** *coarse-fraction particles* are 2.5 to 10 microns in diameter.
- **PM$_{2.5}$** *fine-fraction particles* are 2.5 microns or less in diameter.
- **UFP** *ultrafine particles* are less than 0.1 micron in diameter.

Fine- and coarse-fraction particles differ in origin, composition, and manner of interaction with the atmosphere (particularly with respect to distance traveled). They also differ in their interactions with human tissue.

Coarse-fraction particles result from industrial processes that involve crushing or grinding, such as construction, farming, and mining activities. Coarse-fraction particles in the air typically deposit upon the ground after traveling from several meters to tens of kilometers from the emission source.

Fine-fraction particles originate from combustion sources and from gaseous precursors, such as sulfur dioxide, nitrogen oxides, and certain organic compounds. Foremost among these processes is combustion of fossil fuels, which is considered to be the application of heat that enables carbon and hydrogen to react with oxygen in the air. This releases energy, water vapor, carbon dioxide, and other compounds. The other widespread process to which fine-fraction particles are attributed is high-temperature manufacturing, such as smelting and steel production. Fine-fraction particles can remain in the atmosphere for up to weeks and travel thousands of kilometers before depositing on the ground.[9] Thus, these particles pose a health risk to unsuspecting populations at great distances from the origin of the pollution.

The size of the particles classified as PM in part determines their risk to health. Particles of size 2 to 3 microns tend to deposit deeply in the lungs in the terminal bronchioles and alveoli. Larger particles tend to deposit in the upper bronchi, thus causing different symptoms and disease patterns. The upper bronchi comprise the portion of the respiratory tract that has fewer (compared to deeper in the lungs) interfaces with the blood supply, so upper bronchial symptoms mostly reflect inflammation and include cough, shortness of breath, airway spasms, and asthma-like symptoms. The tiniest ultrafine particles penetrate the deepest parts of the lungs, where they not only can do great direct tissue harm through airway swelling and fluid buildup but also are able to enter the bloodstream.[10] This allows access to the rest of the body, including the heart, liver, and brain.

Symptoms caused by acute (sudden) exposure to PM include shortness of breath, cough, and worsened asthma. Chronic (long-term) exposure can lead to persistent bronchitis, decreased lung function, aggravated asthma, and abnormal heart rhythms. Exposure to $PM_{2.5}$ has been associated with premature death in people with preexisting heart or lung disease, irregular heart rhythms, heart attacks, or asthma flares.[11] Exposures tend to cluster in urban environments in close proximity to vehicular traffic and/or industrial areas.

BREATHING GROUND ZERO

The 2001 terrorist attack on the World Trade Center in New York City offers tragic examples of the potential of PM air pollution to vastly disrupt human health. Many rescue workers and cleanup crews performed their tasks for weeks with inadequate or no respiratory protection, such as masks with filters. Therefore, the workers spent thousands of hours inhaling unfiltered air laden with a significant amount of PM. A recent study concluded that approximately 70 percent of ten thousand workers tested over the two-year period from 2002 to 2004 cited substantially worsened respiratory problems, sinusitis, and gastrointestinal ailments.[12] One-third of the workers demonstrated diminished lung capacity, and roughly one-quarter of the nonsmokers showed breathing impairment. These findings are more than double the rates that would be expected in the general population.[13]

OZONE

Not all components of air pollution are entirely man-made. They can result from exacerbation of naturally occurring phenomena. Ozone (O_3) is naturally formed in the upper atmosphere when ultraviolet radiation comes in contact with environmental O_2 and changes its configuration. This process is essential for human well-being because ozone serves as an atmospheric filter against cancer-causing ultraviolet radiation released by the sun.

However, ground ozone is detrimental to humans, and made even more so by a warmer world. Ground ozone forms when sunlight causes a photochemical reaction of O_2 with ground pollutants, such as nitrogen oxides (see description below) or chemicals released by volatile organic compounds (VOCs). VOCs release interactive gaseous particles far more profusely at higher temperatures. Automobiles, power plants, and other fossil fuel-emitting industries provide the source material for ozone in the form of nitrous oxides and these VOCs. When exposed to sunlight, these compounds interact with O_2 to undergo a temperature-sensitive photochemical transformation that forms ozone. Ground-level ozone coalesces to become a major component of the hazardous air pollution we know as "smog."

The effects of elevated temperature alone upon the generation of ground-level ozone, absent any consideration of precursor emission, is predicted to increase summertime ground ozone in urban regions by approximately one to ten parts per billion over the next several decades, with the largest impact occurring in populated areas during heavily polluted times.[14] To put this increase in context, it's estimated that the current global tropospheric ozone concentration is double that of preindustrial times, and it is expected to increase 50 percent more by 2020.[15]

Excessive ground ozone has been linked to detrimental human health effects. It can cause inflamed lung tissue and result in breathing discomfort, exacerbate asthma, and worsen chronic lung disease. Absorption into the skin can cause local injury, such as itching and swelling, and systemic exposure through prolonged inhalation can inhibit the immune system. Asthmatics or persons with chronic obstructive pulmonary disease are more susceptible than the general population to ozone-related health problems, with a higher risk for reduced lung function and increased sensitivity to allergens.[16] Some studies have shown increased heart disease and cancer risk for persons with chronic ozone exposure.[17] In industrialized regions with warm weather climates, ambient ground ozone levels can exceed recommended allowable exposures for several hours each day.[18] When this occurs, people engaged in outdoor activities, such as landscaping, construction, and recreation, will confront additional risk. In a warmer world, these exposures will increase. When smog is bad, the effects on a population also include lost workdays, greater school absenteeism, and increased emergency department visits and hospital admissions.[19,20] One elegant observation documented the effect of what happens when the baseline ozone exposure of a city is reduced. During the Olympic Games in Atlanta in 1996, automobile traffic was drastically curtailed. The results gleaned from multiple medical databases from the state of Georgia showed that peak daily ozone levels dropped by 28 percent; pediatric emergency department asthma visits decreased by 11 percent; and acute asthma exacerbations declined by more than 40 percent.[21] One logical explanation for the health improvement was the contribution of the decrease in ozone levels.

Long-term ozone exposures significantly affect life expectancy.[22] One health assessment uncovered that roughly 160 million Americans live in areas with unhealthy levels of summertime ozone.[23] Another recent assessment found that ozone levels still exceed EPA standards in many American cities, resulting in avoidable adverse health consequences. These include as many as 2,500 ozone-related premature deaths, three million cases of acute respiratory symptoms, and one million lost school days that could have been avoided each year if National Ambient Air Quality Standards of seventy-five parts per billion of ozone had been met.[24]

DONORA, PENNSYLVANIA, 1948

One of the most significant tragedies in the United States that acted as a catalyst for clean air regulation took place in Donora, Pennsylvania, on the morning of October 27, 1948. A *thermal inversion* caused an air pocket of warm, stagnant industrial emissions to settle into the valley floor, enveloping the hamlet. For five days, a thick layer of sulfur dioxide and nitrogen dioxide suffocated the steel-producing town of fourteen thousand people. By the time a rainstorm finally cleared the smog, twenty residents were dead and roughly half of the population had suffered respiratory symptoms. One of the worst pollution disasters in American history, this event often is cited as the beginning of the clean air movement, which eventually led to the Clean Air Act of 1970. Today's rapid industrialization in areas of great population density around the world makes this cautionary tale extremely relevant.

A CLOSER LOOK AT OTHER FOSSIL FUEL BY-PRODUCTS

Although the following exposures are not direct health effects of climate change per se, they represent potential toxicity from the carbon-based energy consumption that has driven anthropogenic warming. Some gasses, such as nitrous oxide, are the precursors of ozone, explained above. Other gasses, such as carbon monoxide, are on the rise from increased industrialization and automobile use, as well as from wildfires caused by extreme weather conditions.

SILENT BUT DEADLY

Snowflakes sparkled in the light that shone out the window of the rustic cabin. It was crisp and cold, and as the rainwater turned to lacy ice, it fell onto the cold ground unimpeded and intact. Had someone inspected the crystals, she would have seen perfect symmetrical patterns, no two alike. Brisk gusts blew snow sideways against the cabin walls, windows, and doors, so that it was sealed from the outside by a frosting of icy caulk. That was fine with the inhabitants, because the warmth generated by the relic kerosene-burning stoves was otherwise barely sufficient to heat the interior.

The four young girls and their parents settled in for the night, snug under their comforters and looking forward to a good night's sleep before their cross-country ski adventure the next morning. Dad had a mild headache, which he attributed to the altitude of the Colorado Rockies. One of his daughters had complained of slight nausea, which he assigned the same attribution. Everyone was extremely tired, but that was to be expected after a long day's travel from the East Coast.

He awoke with a splitting headache, but if he had been in a deeper slumber could just have easily slept through it and lapsed into unconsciousness. He'd suffered migraines in the past, but this was different. His whole head felt like it was going to explode. He stumbled out of the cabin and vomited into a snow-bank. He was uncoordinated, like he was drunk, but he hadn't been drinking. Something was dreadfully wrong and he knew it. Was he having a stroke? Back inside, he tripped over a chair and fell onto a nightstand, bashing his knee. He shouted in pain, but no one awakened. His children were not deep sleepers, yet they remained motionless in their beds. What was going on? He was an out-doorsman and experienced camper, and so it dawned on his foggy brain. He and his family had somehow been poisoned by carbon monoxide. In the enclosed cabin space without adequate ventilation, the invisible killer had accumulated insidiously. He was impaired, but lucid enough to recall that seconds mattered. Without fresh air carrying oxygen, his wife and children could perish. His wife and four daughters were on the verge of suffering brain damage or dying if they weren't rescued. They weren't just asleep. They had fainted and were uncon-scious, and their breathing was becoming erratic.

Dad became a hero. The few good breaths he drew outside reduced his headache from a "10" to an "8." That kind of head splitter would have put most persons on their knees, but he knew what he had to do. Like a rescuing fireman racing into a burning building, he carried or dragged his family one person at a time outside and lay them on the cold ground. He then bundled them with blankets and moved from girl to girl. After ten minutes, they were all aroused, complaining of headaches and disorientation, each vomiting in turn, but steadily improving. When they were well enough to stumble under their own power, he loaded them into the family van and drove everyone to the hospital.

Certain airborne gasses have the ability to become poisons in our blood and suffocate our cells. A common example of this is carbon monoxide, which is a colorless and odorless gas, classically described as the product of incomplete combustion. Common sources are car exhaust, heating units, and burning from sources as small as cookstoves or as large as wildfires. To understand the pathophysiology caused by carbon monoxide, we first recognize the concept of *gas exchange*, in which the body inhales ambient air containing life-sustaining oxygen, transfers this oxygen to the blood that passes through one functional part of the lungs, delivers the oxygen to the body's tissues via the arterial circulation in exchange for the waste product carbon dioxide, then sheds the carbon dioxide in another functional part of the lungs so that it can be exhaled. Blood is in part made up of red blood cells that contain *hemoglobin*, a transport protein that allows them to carry oxygen and transport it throughout the body.

Inhaled carbon monoxide (CO) attaches to red blood cells at the oxygen-binding site on hemoglobin. Carbon monoxide forms a much stronger bond with hemoglobin than does oxygen and thus can displace it, causing "suffocation" at the cellular level. Acute exposure (often in enclosed or poorly ventilated spaces) to carbon monoxide typically results in influenza-like symptoms of energy loss, fatigue, headache, nausea and vomiting, shortness of breath, and/or heart palpitations. Severe poisoning can cause altered mental status (including coma) and/or heart and circulatory collapse. Chronic exposure to carbon monoxide may cause decreased heart and lung function, as well as nervous system dysfunction, such as decreases in manual dexterity and reasoning. Persons with less physiological reserve, such as sufferers of heart and lung disease, or those with greater metabolic demands, such as pregnant women, are at increased health risk when exposed to carbon monoxide.

Although less deadly than carbon monoxide, two oxygen-containing compound gasses that contribute significantly to air pollution worldwide and are likewise toxic to humans are sulfur dioxide (SO_2) and nitrogen oxides, including nitrogen dioxide (NO_2). Sulfur dioxide, a colorless gas with a noxious odor, results from burning oil or gas fossil fuels, as well as from industrial activities, such as smelting sulfur-containing mineral ores. Because of its origins in fossil fuel combustion, sulfur dioxide often is found in combination with particulate matter and acid aerosols, such as sulfuric acid. Exposure to sulfur dioxide causes respiratory tract irritation and eye inflammation, resulting in cough, excessive mucus secretion, and aggravation of asthma and chronic bronchitis. Several studies showed that hospital admissions due to heart disease and mortality (death rates) increase on days with higher ambient atmospheric sulfur dioxide levels.[25,26] Chronic exposure to sulfur dioxide has been implicated in diminishing lung function.[27]

Persons who suffer from asthma and are active outdoors during times of high sulfur dioxide levels face increased risk for worse wheezing and other respiratory problems.[28] Likewise, children who live in heavily industrialized areas may experience respiratory problems and, like adults, develop chronic breathing disorders. These children also are more likely to make frequent use of emergency departments, in part because they are more prone to lung infections.

Nitrogen oxides derive from combustion processes, such as those used in engines, heating, and power generation. Nitric oxide (NO) and nitrogen dioxide (NO_2) together are known as NO_x and often pronounced "nox." In the atmosphere, NO_x constitute major portions of particulate matter and contribute to formation of ground ozone when exposed to ultraviolet light and warm temperatures. NO_x have been implicated in the health effects of global warming as a precursor for ground-level ozone, but they can cause considerable public health effects in their own right. NO_x irritate eyes, nose, throat, and lungs, causing shortness of breath, nausea, and fatigue. Nitrogen dioxide also has the potential to turn into an acid form when it combines with water in the lungs. As such, it may oxidize, thereby changing its chemical configuration and thus its toxic properties, and cause damage to lung surfaces. It may also react with skin tissue, causing local fluid accumulation and resulting in swelling (edema) and inflammation. Epidemiological studies have linked asthmatic bronchitis and diminished lung function to long-term exposure to nitrogen dioxide at concentrations found in major urban areas in industrialized cities.[29]

SUFFOCATING SUMMER

It had been a brutal summer—hazy, hot, and humid for an entire month without a single day's respite. Every day, fire hydrants flowed in the South Bronx, providing more of a distraction than real relief from the oppressive heat. Sandra lived in an apartment with her family and spent most of her time looking after her younger siblings. The heat made her job much more difficult. The air conditioner was her only way to get through the heat of the day, because the weather was a killer with respect to her asthma. She tolerated cold weather much better. Over the winter she'd been able to string together six months without needing to visit the ER, but in the past month she'd been there three times, each time spending a few hours in the "asthma room"—a surreal, mist-filled lounge with big comfy chairs adjacent to buzzing nebulizer wall units. She was a regular and got great treatment from the doctors and nurses.

She dreaded lunchtime. The kids were always hungry, and more days than not, she had to hike to the market. She grabbed her albuterol inhaler. Thank goodness for the wonder drug that dilated the small airways of her lungs and usually forestalled the uncomfortable wheezing of asthma. She knew that she would need medication today. The TV weatherman reported that the heat index would be topping 110°F (43.3°C). She set out on her chores and was halfway there when she decided to take a puff of her medication. The inhaler seemed less full than normal, and an immediate sense of dread kicked in. The device was nearly empty.

She would definitely need to get a refill on the way back from the supermarket. Sandra rushed her shopping because even the store was hot.

As soon as she started home, her asthma came on fast and furious. The high-pitched wheezing announced that her airways had begun to spasm relentlessly. Sweaty, dizzy, and hungry for air, she felt faint and sat down on the pavement. Then she had to lie down, just before she blacked out. A bystander raced over to help. Traffic stopped. She vaguely remembered lights and sirens.

The paramedics moved swiftly and skillfully. They lifted her onto a stretcher and then rushed her into the vehicle so that they could immediately get her into an air-conditioned environment and place a mask over her mouth and nose. It hissed oxygen and albuterol mist. They raced "red lights and siren" to the hospital. She was far enough along in her asthma attack that the medication wasn't helping very much. The medics prayed for a clear path through the traffic, because they didn't want to have to "tube" her. Minutes later, she was rushed into the ER, past the less-sick patients and straight into the "code blue" area where a full medical team was waiting. A medical technician tried to stick heart monitor pads on Sandra's chest, but her skin was too sweaty for them to adhere. She was no longer able to speak and was barely awake. She could barely communicate by nodding "Yes" or "No." Her airways had narrowed so much that she couldn't move enough air to muster a wheezing sound, let alone talk.

The monitors alarmed relentlessly. She was suffocating. The team moved swiftly to administer a "kitchen sink" of drugs: methylprednisolone to break the inflammatory cascade; ipratropium bromide to reduce secretions and diminish smooth muscle stimulation; magnesium sulfate to augment lung function; and a special formulation of breathable air enriched with helium to cause an increase in ventilation.

At one point, Sandra remembered seeing a doctor preparing the airway intubation equipment. That's how precarious her situation was, but the doctors battled her asthma successfully. Critical minutes were finessed into an hour by the medical team. Three hours later, Sandra was moving enough air into her lungs to hear herself wheeze again. The nurse told her that she would be admitted to the hospital overnight. Her grandmother came in with her siblings. Grandma attempted a stern look, then burst into tears. Sixty years ago, her baby sister, who looked just like Sandra, had died of asthma.

Fly into Denver, Salt Lake City, or Los Angeles and you can see the air quality for yourself. We are long past the days of taking clean air for granted. "Spare the air days" are announced when the smog is so bad that walking outside is hazardous to your health. That situation simply cannot be allowed to persist. Nature did not create this situation. We did. If we keep fouling the air, we all will be forced to wear face masks or stay indoors with the windows closed.

· 8 ·

Water Security

$\mathcal{F}$rom the dawn of man, all humans have shared the essential common need of access to potable water, which is necessary for survival. The link between clean water and human well-being is the most fundamental relationship in public health. Without clean water, infectious diseases and shortened life spans are common. Populations displaced by adverse weather events and conflicts regularly are without access to essential sources of water, often immediately and then for prolonged periods of time. This creates a profound health hazard.

Earth's growing population is a tightening vise on available clean water. Agricultural uses currently account for approximately 70 percent of the world's freshwater use.[1] The sources of our drinking water are affected by climate change, which has imposed numerous stressors on vast watersheds that determine the nature of many of the world's great rivers. Because of water diversion and consumption, in many high-productivity grain-growing areas, rivers increasingly do not reach the sea. Furthermore, the remaining rivers frequently are polluted by industrial wastes and sewage.

More than 1.6 billion people (greater than five times the population of the United States) worldwide do not have access to clean water, 2.6 billion people don't have access to sanitation, and 1.4 billion people live in areas where water use exceeds replenishment.[2] Diseases from contaminated water are the second-highest cause of death in children.[3]

For poor countries that routinely face water insecurity, climate change will exhaust existing resources further and make water access more difficult or even impossible. Much of the developing world will be adversely impacted by more severe and frequent droughts. Droughts lead to extreme dry land conditions, setting the stage for floods. Flooding does not improve access to drinking water, but rather, causes devastation. No country will be immune.

Table 8.1. Climate Change Effect

Climate Change Effect	Impact on Water Security
Air temperature increase	Reduction in glacier-fed reservoirs
Surface water temperature increase	Greater risk for contamination: growth of algae, decreased oxygen content, and less self-purification
Sea-level rise	Saltwater contamination of coastal reservoirs
Shifts in rain patterns	Less-reliable water sources—droughts, floods
Greater seasonal rain variability	More difficult flood control and reservoir use
Increased evaporation from warmer climate	Reduction in water availability, salinization of water, lower groundwater levels
More frequent and intense storm events	Loss of drinking water integrity, risk of pollutant mixing, greater risk of droughts

Source: Intergovernmental Panel on Climate Change: Climate Change and Water, http://ipcc.ch/pdf/technical-papers/climate-change-water-en.pdf. Accessed December 27, 2016.

Climate change will introduce water security challenges into countries that traditionally have had reliable water supplies and few, if any, sudden reductions in their water supply ("water shocks").

The total effects on our water ecosystems have been predicted to be widespread and extensive: accelerated glacier melting; changes in precipitation, runoff, and groundwater recharge patterns; extreme droughts and floods; water quality changes; and saltwater contamination of coastal aquifers. The combination of droughts, storms, and floods will contaminate or wipe out entire water supplies. In the absence of mass human migration, people will be forced to rely upon drinking from contaminated water supplies.

WATER WARS?

Given that 60 percent of the world's population lives in water basins shared by two or more nations, many historians, scientists, and political experts have speculated that the twenty-first century will be one of worldwide conflicts over water.[4] To make matters worse, overpopulation and industrial development are increasing water scarcity worldwide. Shrinking water sources in Central Asia (Aral Sea) and sub-Saharan Africa (Lake Chad) are cited as potential hydro–conflict zones.[2] Water sources in locales with a history of border warfare, such as the mountainous regions separating India and China, and along the Jordan River that straddles Lebanon, Israel, and Syria, risk both new and further exacerbation of current conflicts.

Syria's devastating civil war began in 2011 and has witnessed 300,000 human fatalities, displaced at least 7.6 million people, and created an additional 4.2 million refugees.[5] Although the causes of the war are complex, a key factor was the nation's devastating drought that began in 1998. It caused the most severe set of crop failures in that nation's recorded history and sparked millions to migrate from the rural countryside into urban areas, kindling the strife for future conflict.[6] This drought was almost certainly Syria's worst in the past five hundred years and likely the worst in nearly a millennium.[7] According to NOAA, decreased winter rainfall in the Mediterranean region leading up to 1998 has been related causally to climate change.[8]

The Middle East is a perfect cauldron for the "threat multiplier" of climate change: poor governance, civil strife, and limited economies susceptible to sudden reductions in oil production ("oil shocks"). Climate change is not the sole cause for these refugee crises and wars, but it makes things worse.

Increasing crises lead to human migration because lack of resources and civil conflict drive people from their homes. As climate change predictions come to pass, natural disasters will further destroy precious water resources. Consolidation of large groups of people into small areas, be they refugee camps or whatever spaces are available after natural disasters, further strains limited water resources. These series of events will set the stage for a multitude of infectious diseases to inflict severe damage on an already vulnerable population. The main vector of disease is water contamination.

By far the greatest water-related risk to human health is infectious disease caused by waterborne pathogens: bacteria, viruses, and parasites. The majority of these infectious agents travel in the environment via human and animal feces. In areas where water is scarce, drinking water and latrines for human or animal waste often are comingled. It doesn't take much to contaminate a water supply. One gram of feces can contain ten million viruses, a million bacteria, a thousand parasite cysts, or a hundred worm eggs.[9]

Extreme weather from climate change threatens to magnify this situation. Heavy rainfall can easily exceed and overwhelm the most robust sewage treatment systems. Flooding from heavy rains has been linked to cholera, hepatitis A, and enterovirus infection, sometimes of epidemic nature.[10,11] After a storm, the standing water is a perfect germ breeding ground that easily supports proliferation of the infectious agents that cause common and uncommon illnesses, such as acute debilitating diarrhea, or even leptospirosis or polio.[12] Drinking water contaminations from disease-causing coliform bacteria likewise have been linked to rainfall events.[13] On the other end of the extreme weather spectrum, droughts act as concentrators for waterborne pathogens, because previously safe levels of "dilute" bacteria reach toxic levels, and bodies of water shrink under drier skies and warmer temperatures.

DIARRHEA

Anyone who has visited a developing nation understands the perils of water pollution. When standards for clean water are not well maintained, contaminated water sources frequently become the cause of dreaded "traveler's diarrhea." Affected individuals can do little more between dashes to the bathroom than curl up while suffering abdominal cramping and nausea. For a moment, imagine if diarrhea affected hundreds of people simultaneously in a remote village without a bathroom, where everyone had no alternative but to use a common pit latrine. Furthermore, think about the situation if the community well produced the only available drinking water, which generated all of the suffering victims. Until the water could be disinfected, diarrheal sickness would be an ever-present threat, rendering symptomatic victims unable to work or care for families.

In most circumstances, the severity of disease links to the number of viable infectious organisms that reach the intestines. Disease can strike in many different manners, including intractable vomiting, shaking chills, or total body toxicity that shuts down vital organs. The most common cause of death is from dehydration. For all the complexity of the human body, this situation is easy to comprehend. If a person loses more fluid and electrolytes through diarrhea than can be replenished (orally or intravenously), dehydration worsens. At a certain point, without lifesaving proper fluid and electrolyte replacement, dehydration kills.

No better example of this exists than the scourge of cholera. The onset of symptoms is an insidious process. At first, the victim may suffer nothing more than several days of loose, watery stools. As the disease progresses, symptoms worsen with severe bouts of cramping and profuse watery diarrhea, often striking in the middle of the night and early morning. A sense of unease and depression overcomes sufferers as they experience the onset of dehydration. Total body water deficit robs the body's cells of the fluid needed to perform normal functioning. Without fluid resuscitation, the kidneys begin to fail. The heart begins to pump faster to compensate for lack of blood volume. In hotter climates, sweat glands cannot function properly, which leads to overheating.

EVERYTHING IN THE TIME OF CHOLERA

On Monday, the global weather radar showed a tranquil atmosphere. On Tuesday, early signs predicted a developing "system." By Wednesday, it was obvious that meteorological chaos would be in force within days. The typhoons and hurricanes could be predicted, but not the simultaneous earthquakes, floods, and mudslides. Whether by coincidence or as a feature of a changing climate, the natural events overwhelmed the capacity of governments, the world's health organizations, and nongovernmental relief groups to manage the catastrophes and predictable infectious disease sequelae.

Ian was on the ground in the Dominican Republic treating broken bones and crushed limbs after the massive earthquake centered underneath Santo Domingo. The news media reported the gale-force winds and waves that pounded the Philippines, and yet another tidal wave that struck Indonesia. Naval fleets and hospital ships that would have provided lifesaving services in normal times had been deployed to conflict zones to support military missions in the Middle East and North Korea, so they were not available to assist the millions of people imperiled by the onslaught of natural disasters. Countries fighting each other had few resources to call upon in a time of increasing global non-conflict weather mayhem.

When the good doctor treated the first victim who had all the classic symptoms of the dreaded epidemic diarrheal disease, he knew that he was about to fight a losing battle with what would likely become a pandemic. Like many other survivors in the earthquake zone, Maria lived in a tented camp and suffered diarrhea. However, what was different was not subtle. Within one day of her initial abdominal pain, cramping, and vomiting, she suffered the explosive appearance of voluminous "rice water" diarrhea, more than a liter an hour, that rapidly caused profound dehydration, weakness, and low blood pressure, leading to true physiological shock. The culprit was cholera, an often-lethal diarrheal disease in adults and children caused by the toxin created by *Vibrio cholerae* O1, which colonizes and multiplies within the gastrointestinal tract of the victim. It is spread by contaminated water and food supplies in areas of poor sanitation. These conditions are a feature of natural disasters, such as earthquakes and floods, and also where people are crowded, such as refugee situations.

Given the overcrowding of the refugee camp and lack of medical personnel and resources, Maria suffered for two days before she was prioritized for transport to a location where she might receive aggressive hydration. She died en route, and became one of the 50 percent of untreated victims who dies.

FIFTY YEARS OF CHOLERA

We are in the midst of the seventh pandemic caused by *Vibrio cholerae* O1.[14] This bacterium has ravaged the world at intervals, notably appearing during humanitarian crises that follow natural disasters, during civil strife, and in countries with poor governance that does not mitigate the risks. Whereas previous strains of cholera have burned out after periods of five to twenty years, the current strain, known as *El Tor*, has not only persisted but undergone "hybridization" to become more lethal, killing 1 to 5 percent of victims, compared with less than 1 percent historically.[15] Climate change will act as a threat multiplier to the conditions that allow cholera to thrive, causing the disease to remain a scourge for the foreseeable future.

To grasp the scope of infectious diarrhea, it's helpful to consider how *V. cholerae* and other pathogens, including bacteria, viruses, protozoa, and algae, interact with the human body. These microorganisms use us to thrive and reproduce. Most of the time, the body is effective at protection, either by blocking invasion or interrupting reproduction. In the gastrointestinal (GI) tract, the immune system functions to eliminate invaders, while other "healthy bacteria" (commensal organisms within the GI tract) have claimed most of the available spaces. Similar protective mechanisms guard germ entry through the skin.

During times of good health, our body "neighborhoods" are well regulated and do a fine job of keeping out invaders. But the situation really breaks down when a stressed human physiology is assaulted by forces of extreme weather or a natural disaster. When a person's body is weakened by malnutrition or concurrent illness, vulnerability to infections increases.

Most acute exposures to waterborne pathogens result in the familiar symptoms of gastroenteritis, which may include nausea, repetitive vomiting, watery diarrhea, abdominal cramps, headache, diffuse muscle aching, weakness, and fever. Depending on the germ and magnitude of exposure, symptoms begin anywhere from hours to weeks after exposure. Most people associate illness with a particular food they ate, but much more commonly, it was the dirty utensil, the contaminated ice in the water glass, the salad, or even one's "clean" hands. Fortunately, for most sufferers, these illnesses are self-limited, with the majority of symptoms lasting less than a week and resolving without medication. However, a more severe bout, a disease caused by a more dangerous pathogen, or repetitive episodes can be rapidly debilitating and lead to collapse and death. Certain chronic diseases weaken the immune system and cause susceptibility to infectious diarrhea. Small children and elders are especially vulnerable to worse outcomes. In the former, each episode of diarrhea may jeopardize normal healthy growth by reducing appetite and thereby ingestion of essential calories and nutrients. In children and older persons, a raging infection, typified by typhoid, very rapidly progresses from diarrhea to systemic illness, with substantial death rates.[16]

Let's briefly consider the categories of primary causes of infectious diarrhea: bacteria, viruses, and parasites.

Bacteria

Bacteria are single-celled microbes that inhabit nearly all environments on Earth. Most do not pose a threat to human health and, indeed, some species inhabit the human "microbiome" and are integral to human health. Enteric bacteria, such as *Escherichia coli* (*E. coli*), *Salmonella* species, *Shigella* species, and *Vibrio cholerae*, comprise the most common waterborne pathogens that adversely impact human health. These bacteria jump the moats of our defenses by a simple strategy: carriage in feces, which fuels the fecal–oral route.

Severe gastrointestinal infections may result in dehydration or may spread via the bloodstream to cause diseases such as bacterial meningitis or overwhelming systemic infection that lead to vital organ failure (sepsis). This may cause death, especially in areas with limited resources. Certain bacteria contain toxins that elicit an inflammatory immune response from the body. This phenomenon may cause more disease than the direct toxic effects of the bacterial infection. For instance, a strain of *E. coli* known as O157:H7 has a particular association with hemolytic uremic syndrome, an inflammatory immune response characterized by kidney failure, red blood cell destruction and anemia, and decrease in platelets, which are essential for blood to be able to clot.

Not all bacteria simply come and go in self-limited gastrointestinal infections. Some bacteria have distinct characteristics and generate identifiable clinical syndromes. From a climate and health perspective, all of the following illnesses in this section have been linked to outbreaks caused by rainfall that overwhelms sewage and/or mixes drinking water with agricultural areas, and most have been linked to elevations in environmental temperatures.[17] As temperatures continue to rise, it is likely that these clinical syndromes will continue to impact human health.

MONTEZUMA'S REVENGE

Sherry prided herself on having a cast iron stomach. She would brag to her friends, "I can eat anything, anywhere, any time. I never get sick." She always opined to her travel companions that it was rude to turn down food when they were guests in the home of a family in a foreign country. So, she saw no harm in eating a small piece of dried nak cheese offered by a village elder in a small village in Nepal. She had already partaken of tea with fermented butter and quail eggs that shone with every color of the rainbow. Certainly her gastrointestinal tract had been habituated to the pathogens that might cause acute diarrheal illness. She chewed on the rubbery cheese nugget for fifteen minutes before it became soft enough to be swallowed. She gulped and chased it with a large swig of water from a ladle handed to her by the soiled hands of a well-intentioned yak herder, who previously in the day had collected animal excrement that he manually shaped into plate-sized discs and slapped onto a rock wall to be dried into fuel. Four hours later, Sherry's cast iron stomach was a churning cauldron of microbes that had her racing to the "blue tent," where she and other similarly stricken victims quickly filled the hole with inestimable misery. She was suffering from traveler's diarrhea,

> the most important travel-related illness in terms of frequency and economic impact. The specific bacterial pathogen that got her could have been one of many, but in this case was *Campylobacter jejuni*, a nasty bug by any measure.

Campylobacteriosis, caused by *Campylobacter jejuni* and other species of the genus, is one of the most common diarrheal bacteria-caused illnesses. It appears as watery (sometimes bloody) diarrhea, fever, abdominal pain, headache, and nausea. Humans become exposed through contaminated food and water, or through exposure to sewage during overflow (rainfall) events. *Salmonella* bacteria also cause a similar illness, mostly through contaminated food or drinking water. Salmonellosis tends to be self-limited in most people, but elderly and immunocompromised people (e.g., those with AIDS or undergoing chemotherapy treatment for cancer) more readily suffer from life-threatening, systemic infections. *Salmonella* proliferates during warmer climatic conditions, and as these increase, so likely will the disease burden.[18]

Vibrio is a genus of bacteria that contains several species, the most notorious being *V. cholerae*, which causes cholera. *V. vulnificus* and *V. parahaemolyticus* are other clinically important species endemic to Europe and the United States. They can infect open wounds and cause disfiguring skin and underlying tissue destruction, or in other cases may cause overwhelming systemic infections. *Vibrio*'s natural habitat is coastal ocean or brackish water, where it thrives in warm temperatures. Rising ocean temperatures have allowed northward spread of these pathogens and the resulting subtropical diseases, one example being when passengers became sick while eating local shellfish during an Alaska cruise in 2004.[19] This has become a worldwide phenomenon. Israeli doctors documented a spike in *Vibrio* infections during a recent summer heat wave. The Baltic Sea, which has had unprecedented warming over the past thirty years, has seen an otherwise unexplained emergence of *Vibrio* species.[20]

Escherichia coli is a bacterium that is part of the normal intestinal ecosystem (flora) of humans. There are six "pathotypes" of *E. coli* that cause diarrheal disease; worldwide, these are the most common cause of bacterial diarrhea in children. Combined with illness caused by *Shigella* (see below), more than five hundred thousand people die annually and tens of millions are disabled.[21] The sheer numbers of people infected worldwide compels us to pay close attention to the effects climate change will have on *E. coli*–generated disease. *E. coli* behave like many of the other germs already discussed. They thrive in a warmer world. One recent literature review (a systematic analysis of many other studies) quantified this relationship: for each 1°C (1.8°F) increase in mean monthly temperature, the risk of *E. coli* diarrheal disease increases by 8 percent.[22]

Shigella causes bacillary dysentery or shigellosis, different names for one of the most common waterborne pathogens. Dysentery denotes a general condition, not a specific bacteria or virus. It may be defined as diarrhea containing blood and mucous, and is often associated with abdominal cramping, fever, rectal pain, and frequent bowel movements. Although many species of bacteria, viruses, and parasites can cause dysentery, various *Shigella* species of bacteria have the largest impact on human health worldwide and, like the germs that cause cholera, are highly infectious.[23] Dysentery is distinguishable from other types of infectious gastroenteritis for two reasons. First, it is often associated with humanitarian crises, such as those in refugee or wartime military camps.[24] Second, certain types of dysentery, like those attributable to the genus *Shigella*, cause high morbidity and mortality, so that left untreated, 10 to 15 percent of sufferers will die.[25,26]

Typhoid still kills 1 percent of its victims, an alarming statistic for such a preventable and treatable disease.[27] Also known as enteric fever, typhoid is an infection caused by the bacterium *Salmonella typhi*. This bacteria is transmitted when people ingest it in food or water contaminated with feces from a "carrier" person. Infection begins with symptoms of gastroenteritis, then spreads through the immune system to other organs and presents as a systemic illness with high fevers, rash, headache, and a paradoxically low heart rate. Worldwide, an estimated twenty-two million cases appear per year, causing two hundred thousand deaths.[28]

Typhoid Mary

Salmonella typhi, like many pathogens, succeeds by infecting humans, who then act as the agents of transmission. Of persons infected with typhoid fever, roughly 2 to 5 percent who have already recovered continue to carry the infection without symptoms.[29] Mary Mallon (1869–1938) was the first person in the United States to be diagnosed as a healthy carrier of these germs. Because she was a food preparer and cook, she infected fifty-three people and caused three deaths by unknowingly contaminating food.[30] "Typhoid Mary" became infamous by vociferously denying any role in the spreading of disease and by refusing to stop working as a cook. She died in 1938 while under quarantine.

Viruses

Viruses are tenacious, adaptable microscopic organisms that require the cells of their host to reproduce. Human viral pathogens generally are passed through air, physical contact, and contaminated water. The waterborne viruses that are the most common threats to human health include echovirus, hepatitis A and E viruses, rotavirus, and norovirus. Outbreaks of disease are often seen within dense populations, particularly in areas affected by poor sanitation and/or a limited clean water supply. As stated above, most infect the gastrointestinal tract in a self-limited manner. However, vomiting and diarrhea can cause dehydration rapidly. As for most pathogens, the greatest danger is to persons at the extremes of age: children with fewer reserves to combat water loss and older adults with preexisting medical conditions.

Norovirus outbreaks have been linked to heavy rainfall and flooding, and can rapidly spread through person-to-person transmission, often within a population in close proximity.[31,32] Norovirus has decimated the health of cruise ship passengers, because the confines of the ship allow the virus to disseminate rapidly, infecting hundreds of passengers in a matter of days.[33,34] Hurricane Katrina offered a larger scale example. Of twenty-four thousand evacuees who relocated to Houston and were treated at the Reliant Park medical clinic, 18 percent of persons reported symptoms of acute gastroenteritis, and 50 percent of tested stool samples from these persons came back positive for norovirus.[35] Not only were the evacuees affected, but the rate of secondary transmission to medical personnel, police officers, and volunteers was high. Although norovirus is not known to kill, imagine the additional psychological trauma of debilitating gastroenteritis to the evacuees, not to mention the logistical hindrance to the volunteers and medical personnel called in to help. Norovirus is the leading cause of gastroenteritis outbreaks in the developed world. If heavy rainfall and flooding increase, the number and severity of outbreaks are likely to increase.

Hepatitis A virus occurs in water supplies contaminated with human feces and leads to liver infection, manifested by fever, abdominal pain, vomiting, and diarrhea. In some cases, sufferers may turn yellow in color, a condition known as jaundice, because the stricken liver is unable to help the body clear accumulation of certain products of metabolism. Unlike other forms of viral hepatitis (hepatitis B, hepatitis C), hepatitis A infection is self-limited and does not induce a chronic disease carrier state. Hepatitis E has a disease course similar to that of hepatitis A, with a terrifying additional feature: one in five pregnant women in the third trimester dies if afflicted by hepatitis E.[36]

Poliomyelitis ("polio") is a viral pathogen spread through contaminated water. The virus infects via the intestines and then the nervous system, where a rare but devastating complication causes permanent paralysis, fortunately in less than 1 percent of infected persons.[37] Some infected persons experience fever, fatigue, headache, and vomiting. Even though most infected persons do not develop symptoms, they remain carriers capable of spreading the virus through feces. Children are most at risk for this disease; vaccination campaigns remain the major strategy for reducing worldwide prevalence.

Parasites

The word *parasite* conjures personality traits of opportunistic manipulators feeding off the success of others. That is correct. Biological parasites that prey upon humans have developed very successful ways of thriving and reproducing. Two major classes of parasites present waterborne threats to humans: protozoa and helminths (worms). Infection by these parasites usually implies long-term or chronic exposure to a contaminated water source. In places of poor water access, climate change hastens erosion of sanitary boundaries, bringing these microscopic threats closer to us.

Protozoa. The name *protozoa* stems from the Greek words for "first" and "animals." Protozoa reside in the kingdom *Protista*, along with their single-celled brethren called *eukaryotes*, or nucleus-containing cells. Unlike some other eukaryotes such as algae and slime molds, protozoa emulate animals more than they do plants or fungi. They are *heterotrophs* that survive and grow by consuming other organisms. Parasitic disease resulting from protozoa ingestion varies in clinical presentation, but usually manifests with diarrhea, nausea, vomiting, cramping abdominal pain, constipation (less common), fever, and/or weight loss.

The two specific protozoa types that exert perhaps the highest impact on human health are *Giardia lamblia* and various species of *Cryptosporidium*, which cause approximately 16,000 and 748,000 annual cases, respectively, in the United States.[38,39] Both protozoa assault humans from sewage contamination of reservoirs and recreational bodies of water, and outbreaks of both are linked to extreme weather events.

Cryptosporidiosis bears responsibility for the greatest number of aquatic-recreational disease outbreaks in the United States.[40] *Cryptosporidium* is ubiquitous and tenacious, and one of few pathogens that are highly resistant to chlorination, which is our major strategy to disinfect water. Chronic infections can be detrimental to childhood development, impairing healthy weight gain and growth. In addition, persons who have serious comorbid diseases, such as HIV infection or tuberculosis, suffer a unique disadvantage, in that chronic *Cryptosporidium* infection can interfere with absorption and therapeutic levels of certain disease-specific drugs, leaving the patients less effectively treated and potentially more contagious.[41]

Giardiasis, also called "beaver fever" because of that animal's role as a vector for *Giardia lamblia*, usually results from sewage contamination of drinking water. The CDC receives reports of roughly sixteen thousand cases annually in the United States, but experts estimate its annual incidence is grossly underreported and place the true incidence at approximately two million.[42,43]

Helminths. Helminths are large, multicellular parasitic worms. Clinical disease is determined by the particular species of worm and its developmental stage within the human host. From a global health perspective, helminths rule the parasite world. Approximately two billion people at any given time are parasitized by helminths.[44]

Ascariasis is caused by the roundworm *Ascaris lumbricoides*. Roughly 1.5 billion people worldwide are infected, making this one of the most common parasitic infections.[45] It spreads by the most efficient route into the human host, namely, fecal-oral contact. The roundworm most often moves into a person when sewage has contaminated the food supply, such as after heavy flooding. Although most people remain asymptomatic when afflicted, respiratory symptoms, including cough, shortness of breath, or wheezing, can occur shortly after migration of larvae from the gastrointestinal tract to the lungs. Later, symptoms of abdominal pain, nausea, and vomiting may occur. Acute infections are treated with one to three days of antibiotic therapy. However, most infections worldwide are in places where such medicines are scarce. If ascariasis is not treated, chronic lung or heart inflammation might ensue.[46]

Strongyloides stercoralis, also known as the threadworm, is a parasitic roundworm found in areas served by water supplies that have fecal contamination. Although most infested persons remain asymptomatic, symptoms may appear, depending upon the worm's life cycle and victim's immune state. The worms initially penetrate skin, causing itching and swelling. As they invade the lungs, symptoms of coughing, wheezing, and shortness of breath may occur. Vague gastrointestinal symptoms of nausea, vomiting, diarrhea, or constipation follow as the worms migrate to the small intestine. Persons who are immunocompromised risk becoming chronically infected or suffering a "hyperinfective" syndrome with unchecked reproduction of the parasite, which can progress to catastrophic clinical manifestations of shock, disseminated intravascular coagulation, meningitis, renal failure, and/or respiratory failure.[47] The hyperinfective syndrome carries a mortality rate of 60 to 80 percent.[48]

Trichuris trichiura, or human whipworm, is a common intestinal helminth. Poor hygiene and limited access to clean water are risk factors for infection. Most infected persons are asymptomatic. Symptoms are limited to the gastrointestinal tract and usually seen only with a heavy parasite burden. Dysentery, anemia, and vague abdominal discomfort are the most commonly cited manifestations. Roughly 25 percent of the world's population is infected with whipworms. This number is expected to increase as climate change modifies the environment.[49]

Hookworms of significance include *Ancylostoma duodenale* and *Necator americanus*. Anemia is a common condition seen with hookworm infection, because the worms ingest red blood cells while they reside in the human gastrointestinal tract.

Guinea worm disease, also known as dracunculiasis, is caused by the roundworm *Dracunculus medinensis*. This genus takes its name from the Latin for "affliction of little dragons." Dracunculiasis is a human-to-human disorder linked to drinking stagnant water. Once ingested into the GI tract, *D. medinensis* larvae penetrate the stomach and intestinal walls and travel through the body to invade soft tissues, usually in the legs. The worms exit by burrowing out through the skin, creating burning and painful blisters that, when exposed to water, release the larvae and recontaminate the water supply. These skin sores are painful, may become infected with bacteria, and often result in disability if they occur near joints. People most at risk live in remote areas of sub-Saharan Africa. Guinea worm disease enjoys a world of humans without a preventive vaccine or medicinal cure. In the mid-1980s, the Carter Center in conjunction with UNICEF began a campaign to eradicate the disease. At that time, twenty African and Asian nations suffered 3.5 million cases.[50]

With implementation of educational initiatives and simple water-filtering technologies, the number of reported cases has decreased vastly, almost to the point of eradication, with the CDC reporting only twenty-two cases in four African nations in 2015.[51] Given the possibility of vast climate alteration from our quickly changing environment, it is possible to conceive that Guinea worm disease could make a comeback in the future, which would once again cause significant morbidity throughout sub-Saharan Africa.

Schistosomiasis is a preventable disease caused by skin-penetrating helminths. This waterborne parasitic infection is caused mainly by three flatworm species: *Schistosoma haematobium*, *S. japonicum*, and *S. mansoni*. Humans are the definitive host, although in Asia certain livestock are important reservoir hosts. Exposure to schistosomiasis occurs by swimming or bathing in water infested with larvae that have the ability to penetrate human skin. This cutaneous invasion causes an itchy rash and fever, and then larvae subsequently infiltrate blood vessels, where they migrate to the liver or bladder. Once in these locations, they begin producing eggs, which can cause an inflammatory response leading to organ damage.

Schistosomiasis is endemic in seventy-six countries, most of which are in Africa, but also in Asia and the Americas (Caribbean).[52,53] The World Health Organization cites that at least seven hundred million people are at risk of infection and two hundred million suffer from schistosomiasis. Of the infected persons, 120 million have symptoms and 20 million have severe disease, with more than two hundred thousand deaths per year in sub-Saharan Africa alone.[54,55] Improved sanitation and water treatment reduce the threat of acquiring schistosomiasis, but as water security deteriorates from climate change, vulnerable nations with poor infrastructure will be most at risk for increased outbreaks leading to continued morbidity and mortality.[56] In China, biology-driven climate models have concluded that weather changes will drive expansion of schistosomiasis into currently non-endemic areas.[57]

Infernal Itching

A few more diseases bear mentioning. Although not waterborne pathogens per se, the World Health Organization acknowledges them as being exacerbated by a lack of available clean water, and thus can only be forecasted to increase in the future as our climate continues to change.

IF IT ITCHES . . .

The water was hopelessly muddy, so nothing could be kept clean. It had been years since the river flowed clear and green. Agricultural runoff laden with fertilizers and the effluents of industry fouled the water. Environmental enforcement was lax. The new norm was water that was too toxic to drink because it was more likely to breed toxic algal blooms than to support lily pads and trout. With brown as the prevailing color, the mentality adapted to dirt under fingernails, dusty eyebrows, and the constant sensation of grit in one's mouth. It wasn't the Dust Bowl of the thirties, but something more insidious and permanent. Hygiene inevitably suffered. Skin diseases became the norm, and it seemed that everyone itched all the time.

One of the parasites that flourished in this landscape of filth was scabies, caused by the human scabies mite *Sarcoptes scabiei* var. *hominis*. These minuscule mites complete their entire life cycle on the skin of humans. Usually acquired during sexual contact, they hop onto people from clothing and bedding.

Andrew first noted severe itching at nighttime between his fingers and on his wrists and elbows. It seemed worse when he sat in front of his fireplace. He didn't know what to make of it until he saw a tangle of very faint red lines, which were serpentine burrows on the surface of his skin created by impregnated adult females depositing eggs. Within days, his scalp was infested, and soon, the entire family was scratching. What he didn't know was that without treatment, the infestation could persist indefinitely. After a few weeks of pruritic misery, he sought treatment. Because scabies was now a common household disorder, the doctor knew what to do. He treated everyone with creams and lotions, but until the affected top layer of skin was shed after an itchy month, they all scratched a lot.

Itchy scabies afflicts populations plagued by poor hygiene and is a hallmark of destitution. A contagious skin infection, it spreads rapidly in crowded conditions, made even more receptive by lack of clean water and hand sanitation. The pathogen is a microscopic mite that has thrived everywhere from the back alleys of London to the causeways of Calcutta. Its chief manifestation is an intensely irritating rash of tiny bumps that are red in color and most commonly found in the webbing of fingers or skinfolds of joints. Its unmistakable manifestation is intense and incessant itching that provokes nearly maddened scratching and excoriation of affected skin.

Tinea infection (ringworm), another itchy malady, is a fungal skin disease that can affect skin, scalp, or nails. It's a misnomer, because no worm is involved. It spreads by direct contact with an infected person or animal, or from items contaminated by the fungus, such as clothing, towels, wrestling mats, and bathroom items. Antifungal medicine can help but is futile in a living situation with limited access to clean water and soap, such as a refugee camp or crowded living conditions that are likely to become more common as climate change and other events force human migration.

OUR MARBLE NEEDS TO STAY BLUE

As any astronaut who has gazed down during spaceflight can tell you, Earth is a planet of water. But only 1 percent of it is potable for humans, and less than that is safe to drink. Population growth will increase water demand at least well into the mid-twenty-first century, and climate change will degrade water security through extreme weather events and sea-level rise. In the majority of countries today, existing water supplies are insufficient to meet urban, industrial, agricultural, and environmental needs. The UN World Water Development Report predicts that if increasing water needs grow on the current trajectory, global demand for water will exceed its supply by 40 percent in 2030.[58] Global health efforts have made tremendous progress in the past few decades to reduce disease from waterborne pathogens. However, the assumption has been that we live for the most part in a steady state with regard to waterborne diseases. If these diseases grow tremendously, all bets are off. Most alteration of water results from human actions. We cannot contaminate and consume water as if we have a limitless supply, because we neither have enough nor have we shown the discipline to enforce its proper use and preservation. We can only hope that the water insecurity that results from climate change does not significantly add to the burden of illness in an increasingly thirsty world. This is an issue of sufficient importance to require constant surveillance, strictly enforced regulations, and strategies to conserve, purify, and protect every water source on this planet.

• 9 •

Food Security

$\mathcal{F}$ood, water, and shelter form the essential triad for human safety and security. It doesn't take imagination to understand that when any of these come into the crosshairs of climate effects, people are the targets. We have already discussed the impact of climate change on the water security of the planet. Food security is also threatened. The path from farm to table is long and vulnerable. As the planet warms and ecosystems change due to extreme weather, many environmental perturbations might weaken a link in the food chain: plant health and agriculture, animal reproduction and growth, fisheries and aquaculture, food trade and distribution, and consumer behavior. There is much to discuss and ponder in this category of security.

Worsening food security can occur in many ways other than declining availability. For example, it can be caused by lowered nutrient content, waxing and waning agricultural supplies, and diminished food use. Each of these factors directly and indirectly affects human health. The direct effect will be worsening undernutrition, which already exists as an enormous burden across the planet. Indirect effects include increased environmental exposures to toxins and pollutants through extreme weather events that disperse these substances, and through weakening of our health-sustaining ecosystem services.

Ecosystem services are comprised of the different ecosystems present on Earth that provide services to sustain life, such as temperate rainforests that buffer the vast amounts of carbon dioxide that humans produce by storing carbon in their trees and soil. While mitigating human influence, these rainforests provide certain goods, such as food and timber. In addition, they provide less obvious services, such as providing an environment for growth and survival of insects and small rodents that control tick populations and thus mitigate risk for certain vector-borne human diseases, such as Lyme disease. A temperate rainforest is just one example; others include temperate plains, coral reefs, and freshwater marshes, all of which provide components necessary for life on Earth. As these ecosystems are challenged by the rapidly changing environment, their survival and the services they provide are threatened.[1] As ecosystem services are threatened, the nutrients and food we derive from them also will be threatened, leading to food insecurity and lack of the resources needed to provide sufficient nutrition in an already nutritionally challenged world.

FEATURES OF FOOD SECURITY

Food security has many features that combine to equal uninterrupted access to nutritious food. Insecurity can occur because people don't have enough money to buy their next meal, or because they live in "food deserts," that is, communities in which a selection of nutritious food is not available. Food security requires stable agricultural productivity and absence of civil strife that would disrupt a functioning economy necessary for food to get to market. To cite Daniel Silverstein,[2] food security adviser to the United States Agency for International Development (USAID), for food security to exist, seven factors need to be in place. Food has to be:

1. available,
2. accessible,
3. affordable,
4. acceptable in the eyes of the beholder,
5. nutritious,
6. safe, and
7. of good quality.

Climate change has the potential to disrupt all of these factors. The earth is a complex system that does not respond uniformly between or within geographies to increased greenhouse gasses or temperature increases. How does this affect food security? There is some evidence to suggest that slight warming (1–3°C [1.8–5.4°F]) and increases in carbon dioxide (the gas that plants absorb to make energy) will have a modest beneficial impact on the major rain-fed crops of maize, wheat, and rice.[3]

However, in tropical regions of the world (which is the location of most developing nations, wherein live the most vulnerable populations), even slight warming will reduce crop yields and negatively impact food production.[4] As we approach the 3°C (5.4°F) increase for the planet, global agricultural yield begins to slide on a slippery slope, because these yields are expected to drop precipitously if the atmosphere closest to the earth's surface warms in excess of this amount of temperature rise.

Not only will food availability become threatened, but so will the nutrient content within the food. Elevated levels of carbon dioxide in the atmosphere will occur at the expense of nitrogen, which is a key ingredient for plant growth. This directly affects protein concentration and vitamin content for most food crops, and thereby undercuts their nutritional value for humans.[5] Food yield from the oceans also will be dramatically stressed. Most of the earth's greenhouse gas increase in carbon dioxide is absorbed by the oceans, directly changing the pH (acid-base status) of the water.[6] Acidification is a major factor in bleaching of the coral reefs across the planet, which signifies impending or actual death of the live corals.[7] With this comes a decline in aquatic biodiversity and instability of the marine food chain. Increased seawater temperatures will continue to undermine the quantity and quality of fisheries. Furthermore, sea-level rise will change the salinity of coastal estuaries and threaten stability of fisheries and the associated production and harvesting of seafood.[8]

FOOD STABILITY AND ACCESS

As extreme weather events (e.g., droughts, floods, erosion, hurricanes, storm surges, forest fires) increase, food supplies will become less predictable and reliable. Unpredictable weather conditions confound farmers' ability to bring their crops to market. Extreme weather events themselves can cause crop failures with a predictable undesirable aftermath. Droughts, floods, and higher temperatures will change the balance of ecosystems, allowing invasive species, such as animal pests, plant weeds, and algae blooms, to proliferate and harm existing agriculture.[9] Such conditions favor fungal species that can overwhelm crops and contaminate animal feedstocks. An example is the fungal strain *Phytophthora infestans*, which was responsible for the great Irish potato famine of the 1840s, and its variation that currently causes $6 billion in crop damage per year in the United States alone.[10,11]

All of these changes will impact food economy; climate variability is expected to impact food prices substantially. Some experts predict up to a 40 percent increase in prices if the global mean temperature rises 3°C (5.4°F).[12,13] If more disposable income needs to be spent to purchase food, families will have less means to generate surplus income; in an unfortunate cycle, everyone will have less ability to obtain nutritious food. Calorie-rich yet nutrient-poor foods will fill the void, but the increased propensity for catastrophic medical problems (see below) still will remain. A very predictable bad outcome is outright hunger accompanied by an insidious state of chronic malnutrition or even starvation.

NUTRITION SECURITY

As the amount, quality, access, and cost of our food deteriorate, health will suffer directly. Climate change will exacerbate the enormous existing burden of undernutrition and the suffering and disease that go with it. For impoverished people living in remote areas, social safety nets do not exist. If they run out of food, they're out of luck. Today in developing countries, approximately one-third of children are underweight or stunted in growth and various forms of development. Undernutrition is the cause of more than one-third of deaths in children younger than five years old.[14]

The socioeconomic costs of undernutrition are profound and cruelly hit women and children the hardest. Undernutrition works hand in hand with infectious diseases, because weakened immune systems are unable to fight off bacterial infections and parasites. If a person believes that the first thousand-day period of a child's existence sets the trajectory for life, as is endorsed by the American Academy of Pediatrics, then it most certainly should not be characterized by chronic starvation and disease.[15,16] If a child is deprived of body-building nutrition during this critical period of development, then stunting of the mind and body will result, trapping the unfortunate child in a vicious negative health cycle.[17] She will endure frequent illnesses, underperform at school, and be unable to muster the capacity to fully contribute to her family, community, and society. This is typical of how poor health reinforces a cruel situation of poverty. Depriving humans of adequate nutrition causes a condition of weakness that can readily become multigenerational.

FOOD USE

Food use describes a person's ability to absorb and use food nutrients. It is intricately related to that individual's baseline health and access to sanitation and clean water. Increased precipitation and flooding contribute to food- and waterborne infectious diarrheal diseases. These diseases directly affect nutrient absorption and food use because a gastrointestinal tract infected with bacteria or parasites directly decreases the body's ability to absorb nutrients from food. As the world faces an increase in parasite infestations, it likely will see a direct increase in nutritional deficiencies of iron, iodine, and vitamin A; this situation already is present in children throughout the developing world.[18] The downstream impact of these nutritional shortfalls is an insidious cascade that will sap human potential. Iron deficiency leads to extreme fatigue as the body becomes unable to produce enough red blood cells to carry oxygen effectively. Iodine deficiency can result in the condition of cretinism, which causes severe physical deformities and stunted physical and intellectual development. Deprivation of Vitamin A can result in stunted growth, night blindness, and even death. As our climate continues to change, we likely will see an increase in parasite infestations that contribute to malabsorption. These infestations have been correlated with impaired physical and cognitive growth in children.[19] Several worm species act on the human body by colonizing the lining of the gastrointestinal tract, where they prevent absorption of nutrients, cause chronic inflammation, induce diarrhea, and promote blood-loss anemia.[20] Worm life cycles have periods of environmental dormancy that render them very susceptible to changes to their residential environment.[21] We are not entirely certain whether they will pose a greater risk with stressed food and water access, but it does not appear that they will diminish in number, risk of exposure, or health implications.

The scientific link between climate change, weather instability, and undernutrition is becoming increasingly clear. Studies have determined that children born in East Africa during a drought were 36 to 50 percent more likely to be malnourished than were those born during non-drought conditions.[22] A study in Niger supported that 62 percent of children born during a drought had stunted growth.[23] Physical displacement of persons, such as occurs with migrations, due to climate events will compound food insecurity, separating the general population and markets from farmers and fishermen and their harvests.

HUNGER

According to the IPCC 4th Assessment Report, it is estimated that by the year 2080, two hundred to six hundred million people will suffer from undesired hunger.[24] Calorie availability in the developing world is likely to decline dramatically by 2050, resulting in an additional twenty-four million undernourished children. That translates to 21 percent more undernourished children than would exist in a world without climate change.[25]

Hunger and food security often are used as synonyms, but they mean different things. Food security means a successful series of events between the farm or fishery to one's stomach. Hunger is a nutritional health issue, and is one of the socioeconomic determinants of health. In other words, hunger is a disease of poverty. In the developed world, no one ever goes truly hungry unless there is a disaster. The hunger that we know may make us irritable and cause us to have a difficult time concentrating. Among the truly hungry, however, there is no crankiness. There is weakness and wasting. The fifty-five million children in the world who are undernourished quickly learn that playing, fussing, or crying demands energy and takes away strength from what little precious nutrition they have.[26] Blank faces with sunken cheeks and eyes regularly show us that real hunger is emotionless.[27]

As climate change erodes food security, it will not only undermine direct access to food and nutrients, but also push people toward *hidden hunger*. This is created by a growing epidemic of calorie-rich, nutrient-poor foods.[28] In developing countries, hidden hunger occurs when populations primarily consume one food type, such as wheat, rice, or corn, for all of their meals. The term pertains to people who can't afford or find nutritious food and are forced to eat "empty" calories just to stave off hunger pangs. To make matters worse, these food types are beginning to have diminishing returns in terms of production and already low nutrient content. Increased carbon dioxide, secondary to climate change, finds its way into the soil and reduces overall mineral concentrations by as much as 8 percent. This results in an increase in the total starch and sugars within the crops, thereby reducing nutritional value and contributing to the prevalence of hidden hunger.[29]

According to the UN Food and Agriculture Organization, by 2050 the world will need 70 percent more food than is currently produced.[30] This is all on a planet with hotter temperatures, less farmland, less access to clean water, and more unpredictable growing seasons than at any other time in human history.

THREAT MULTIPLIER: ENVIRONMENTAL CONTAMINANTS AND CHEMICAL RESIDUES IN THE FOOD CHAIN

During the Dust Bowl of the 1930s, which was a combination of drought and land use mismanagement, President Franklin D. Roosevelt commented that a "nation that destroys its soil, destroys itself." As have air and water, our soil has absorbed the global impact of synthetic processes and their resultant wastes. Such pollution poses obvious and evolving threats to health. The numbers are staggering. Every year in the United States, 2.2 billion pounds of pesticides enter the environment, many of these containing chemicals known to be persistent organic pollutants (POPs).[31] More than one hundred active ingredients in pesticides are on lists of substances that cause cancer (carcinogens).[32]

Predictably, most soil pollution occurs in locations close to industry or population centers, especially those with antiquated infrastructure, where the pollution stems from spillage and/or leeching from factories, storage areas, and sewage systems. There is outright intentional dumping of chemicals. In a stable and predictable environment, we might be able to avoid soil contamination. If we were able to store chemicals safely, avoid areas of known elevated concentrations, outlaw the most toxic substances, and maintain the ecosystems that distribute and break down inadvertent exposures, we could to a certain degree protect our soil. However, an unstable climate will thwart such efforts. Floods, droughts, and other extreme weather events disturb the environment in unpredictable and powerful ways. Like a colossal blender, climate change–fueled extreme weather events add disruptive energy to ecosystems, mixing toxins with our food and water supplies.

The list of injurious substances that humans introduce into the environment is extensive and includes pesticides, petroleum hydrocarbons, heavy metals, and organic solvents.[33] These all have the potential to cause toxic syndromes, birth defects, and cancer. These pollutants are especially dangerous because so many of them do not biodegrade, thus presenting potentially long-term exposures that logically would be predicted to impact human health well beyond the time and perhaps the location of the initial contamination. Furthermore, some of these substances bioaccumulate in plants and animals, threatening the foundation of the food chain. The combined threat of toxic chemicals and an unstable environment migrates from industrial sites to our food supply. Any increase in toxins or perturbation of the environment will worsen the situation. The health impacts are real and costly.

PERSISTENT ORGANIC POLLUTANTS

When we speak of widespread and persistent toxins, we begin with the worst offenders, referred to above as *persistent organic pollutants* (POPs). The vast majority of POPs are human-derived compounds that resist chemical, biological, and photolytic (sunlight) breakdown. They are the residual blight of a bygone industrial era, and these compounds are incredibly stable and dangerous. According to a recent study by the National Oceanic and Atmospheric Administration (NOAA) and the Fish and Wildlife Service, fish sampling in the Hudson River (after years of polychlorinated biphenal [PCB] dumping by two General Electric capacitors, ending in 1977) showed that PCBs won't fall to levels allowing for safe fish consumption for decades longer than originally predicted.[34]

For most of our "routine" domestic garbage, the health threat defuses when we haul the detritus to remote areas and dump it. However, certain toxic chemicals do not degrade over time. When we do not dispose of wastes in a manner appropriate to the time and method of their decomposition, they pose persistent hazards. The term *bioaccumulation* describes the cumulative concentration of substances in a living organism. Bioaccumulation refers to plants or animals that have undergone long-term exposure to concentrated substances, such as the POP dichlorodiphenyltrichloroethane (DDT) or the heavy metal mercury. Substances enter the organism through natural contact with the environment, such as soil or water absorption, food ingestion, skin contact, and breathing. The amount of bioaccumulation in a living creature depends on the substance's chemical characteristics, rate of uptake, changes of the substance that occur within the animal or plant, and rate of elimination (such as excretion). Bioaccumulation can be rapid or insidious. Very low levels of environmental toxins can grow into serious health threats over time. Uptake occurs first in dirt dwellers, such as plants and worms. Next, the toxins move into the fish and birds that consume the plants and crawlers. Small animals next consume and bioaccumulate the toxins. From that point, the toxins concentrate at an accelerated pace in carnivores and omnivores at the top of the food chain.

The threat of bioaccumulation entered our consciousness with publication of Rachel Carson's groundbreaking book *Silent Spring* (1962). This highlighted the toxicity of DDT and certain health threats of POPs: endangered reproductive health, endocrine disorders, diabetes, and cancer. Fortunately, DDT and several other POPs were banned globally in the 1970s, although DDT still is detected routinely in our environment worldwide.[35]

We synthesized the chart below to demonstrate how toxins such as DDT concentrate in higher order species, and accumulate more densely as they move up the food chain through a process biologists refer to as *biomagnification*.[36,37,38,39]

Table 9.1.   DDT in Lake Michigan Ecosystem Strata: 2000

Water	2 parts per trillion
Bottom Mud	15 parts per billion
Phytoplankton	35 parts per billion
Zooplankton	90 parts per billion
Water Fleas	400 parts per billion
Fish (carnivorous)	5,000 parts per billion
Birds (carnivorous)	100,000 parts per billion

Whereas bioaccumulation describes the increase in concentration of a substance in the uptake between the environment and a living organism, biomagnification describes how a substance can increase exponentially in each successive organism up the food chain. These concepts should be understood for any substance that is harmful and enters the environment, either by natural means (e.g., plankton accumulations in ocean waters that cause colored toxin-laden "tides") or artificial means (e.g., lead leaching from antiquated, eroding municipal water pipes).

Toxins

If increased weather energy from climate change agitates the environment, we should expect a recrudescence of latent toxins. Fragile ecosystems, such as the polar regions, are the most susceptible. The *Arctic Climate Impact Assessment* notes that natural water and wind pathways altered by climate change very likely will result in enhanced bioaccumulation of POPs and mercury in fish and other aquatic animals, leading to high dietary exposure in humans.[40] Mercury provides an excellent case study of how climate change will increase an existing environmental toxin's footprint. Warmer ocean temperatures have been demonstrated to increase the methylation of mercury (which creates the form that is toxic to humans) by 3 to 5 percent for every 1 degree Celsius (1.8 degrees Fahrenheit) increase in temperature.[41]

The human body takes up and stores mercury with alarming ease. This elemental toxin has a long history of badly injuring the human nervous system and liver. It doesn't take much. Mercury occurs sparsely in the natural geologic environment, but industrial processes have augmented its presence significantly throughout the world. Certain species of edible fish, such as tuna, bioaccumulate mercury to the extent that it attains levels that are extremely toxic to humans. Fish consumption has become the most common source of mercury exposure in people.[42]

Mercury exists in the environment in different forms, some of which do not threaten human health until an event, such as extreme weather or a construction project, causes it to move into position to be bioaccumulated in people. Around James Bay in Northern Quebec, Native Americans survived for centuries by eating large quantities of fish containing naturally occurring mercury. After a large hydroelectric dam was built in the late twentieth century, changes in the region's aquatic ecosystems mobilized large amounts of ground mercury into the water supply. Waterborne bacteria changed the environmental mercury into methylmercury, a form that organisms easily absorb but do not easily excrete. The food chain began to bioaccumulate methylmercury, placing the local human population at risk.[43] We can expect extreme weather in the form of heavy precipitation, flooding, landslides, and drought to have the same impact on aquatic ecosystems in this and other locations, and expose greater numbers of humans to toxic forms of mercury.

Mercury poisoning has a well-known toxic clinical medical profile, particularly to the developing nervous system. Research has clarified that the highest at-risk population is children of women who consume large amounts of lake fish and certain seafood species, because of mercury bioaccumulation during pregnancy. The afflicted human develops tremors, mood changes (that can be extreme), and blurred vision. Within the nervous system, mercury blocks normal breakdown of the neurochemical transmitters called catecholamines, sometimes referred to as "fight or flight" hormones. The result is an overstimulation syndrome characterized by sweating, rapid heartbeat, increased salivation, and high blood pressure. Mercury also affects the formation of myelin, a material that covers individual nerves to ensure efficient signal transmission within the nervous system. Dysfunction caused by mercury poisoning presents with muscle weakness, mood disorders, aberrations (e.g., itching and burning) and deficits of sensation, and lack of muscle coordination. Children may have symptoms of red lips and cheeks; loss of teeth, hair, and nails; and skin rashes. They may also develop dramatic, extensive skin peeling from the palms of the hands and soles of the feet.

Pesticides

Pesticides are an integral, necessary part of food security because they allow crop protection and increased food production. However, they can be quite toxic and therefore present a double-edged sword. The most common pesticides used today are organophosphates (malathion, parathion, diazinon, fenthion, dichlorvos, chlorpyrifos, and ethion) and carbamates (methomyl, carbaryl, methiocarb sulfone, butocarboxin, and propoxur). Although over time some have achieved safer health profiles worldwide, their ubiquitous presence continues to pose a threat to human health, particularly in the face of growing demand for food and a changing and unpredictable climate. Current agricultural practices may no longer work to achieve food security in a warmer climate, where insect pest numbers and their range are expected to increase.[44] New agrichemicals and increased use of this category of products is likely, because many modern pesticides have limited effectiveness in dry conditions and degrade faster in higher temperatures.[45] All of these conditions portend greater future use of pesticides. To a certain extent, this will contribute to environmental contamination in part caused by persistent residues on crops.[46,47]

Persons most at risk will be those who have regular, close exposure to pesticides: agricultural workers and populations in close proximity to farmland. Exposures occur not only through direct produce handling, but also via agricultural runoff and from residual chemicals remaining on crops. These chemicals can suspend in air, causing "pesticide drift" to nontarget areas downwind. Some pesticides that are no longer commercially available (e.g., organochlorines, such as DDT) act as persistent organic pollutants (POPs), leading to long-term soil contamination. Fortunately, the side effect profile of organophosphates, which are the pesticides used most commonly worldwide, remains relatively safe. These chemicals have minimal impact on the environment compared to their previous alternatives. However, they still have a detrimental impact on health when humans are exposed to concentrated doses or prolonged contact.

We should consider the clinical impact of climate change in the face of eroding food security. To make up for diminishing yields, farmers worldwide may turn increasingly to pesticides to jump-start food production and maintain bountiful harvests. In the coming quest to resuscitate stressed-out and desiccated farmland, one can reasonably predict more toxic exposures. If this occurs, based on what we know now, patients will present with syndromes that can be characterized as acute or chronic.

Acute exposure will generate patients because of contact to large amounts of toxic pesticides over a fairly short period of time. Such accidental exposures blessedly are rare and mostly limited to agricultural workers who apply these compounds to crops and to persons involved in manufacture and distribution of pesticides. Large amounts of pesticide concentrates can enter the body through contact, inhalation, or ingestion. They block an important enzyme in the nervous system and trigger a cascade of symptoms known as the *cholinergic syndrome*. Physicians use the mnemonic SLUDGE to recall the specific conditions of this type of poisoning:

- **S**alivation: excessive drooling
- **L**acrimation: runny nose and tearing of the eyes
- **U**rination: excessive urge to urinate
- **D**iaphoresis: profuse sweating
- **G**astrointestinal: abdominal cramping and diarrhea
- **E**mesis: vomiting

Other hallmarks of poisoning include pinpoint pupils, asthma-like symptoms, visible muscle twitches and spasms, weakness, and altered mental status, including confusion and sometimes seizures. Organophosphate pesticides have a garlicky odor, and poisoned patients often present to their caregivers exuding a similar odor.

Children comprise a large proportion of acute, accidental pesticide poisonings worldwide. Many play or crawl in areas with recent pesticide application. Their higher ratio of surface area to body size requires a smaller dose for toxic symptoms to appear than that necessary to sicken adults. A younger age at the time of exposure has been shown to confer a higher risk for development of chronic symptoms associated with pesticide exposure.[48]

Pesticides kill pests. They also kill people. The lethality of acute pesticide poisoning is a well-known phenomenon. Easy to obtain and readily available, pesticides can be ingested as a method to commit suicide. This unintended use has risen globally, particularly in rural areas of developing countries. Because treatment depends on supportive medical care and antidotes, which often are not available in the locations of such poisonings, it is a public health issue of significance, with roughly two hundred thousand deaths annually in the developing world.[49]

The second type of pesticide exposure is chronic, involving long-term contact with lower doses of pesticides. Symptoms may be vague and slow to appear. The science here is less clear, but consistently points to a link between long-term exposure and chronic disease. Many scientific studies have linked long-term exposure to afflictions that include birth defects, skin disorders (such as allergic reactions, chloracne, changes in skin color, and in some cases skin cancer), diabetes, cancer, and neurological problems, including memory dysfunction and Parkinson's disease.[50] One study by the National Institutes of Health established a link between diabetes and more than thirty thousand professionals who worked with pesticides for more than one hundred days in their lifetimes.[51] Other studies linked chronic pesticide exposure with conditions such as memory deficit, skin problems such as those listed above, emotional depression, and cancer.[52] Researchers have investigated extensively and established a link between pesticides and non-Hodgkin's lymphoma, as well as certain leukemias, sarcomas, prostate cancer, and multiple myeloma.[53] They have also attributed a cascade of nonspecific symptoms to long-term pesticide exposure.[54] These include weakness, fatigue, headaches, nausea, chest pain, deficits in concentration, and bouts of confusion. Certain reproductive dysfunctions long have been connected to pesticides. These include infertility and birth defects, such as cleft palate, musculoskeletal deformity, and nervous system defects.[55]

Jake Leg

One of the best-documented examples of chronic organophosphate poisoning came out of a Prohibition-era drink. Jamaican ginger extract, known as "Jake," was a patent medicine that contained 70 to 80 percent ethanol. Jake quickly became the medicine of choice for those spirits-deprived persons looking to bypass the restriction on consumption of alcohol. Authorities quickly caught on and forced a change in formulation of the beverage to one that was more bitter in taste and thus less palatable. It was then that a pair of entrepreneurs began to market the drink with an additive, tri-ortho-cresyl phosphate (TOCP), which preserved its legality under Prohibition laws, while making the beverages less bitter and more drinkable. TOCP was an organophosphate. It was later discovered to be neurotoxic, causing weakness and paralysis in limb muscles. The afflicted exhibited a distinctive "Jake Leg" or "Jake Walk." Some blues songs of the early 1930s refer to the Jake Leg, which is the condition now known as organophosphate-induced delayed neuropathy.

Fertilizer

Agriculture requires chemicals not just to kill pests but also to grow a desirable quality and quantity of crops. Phosphorous, nitrogen, and potassium are mainstay ingredients of chemical fertilizers. These replenish soil nutrients to agricultural soil that is in continuous use. These chemicals have proven safe when they are used properly and limited to agricultural areas. However, soil doesn't always behave in fully predictable ways. Weather extremes will move it around and sometimes across barriers. The chemicals will be carried well beyond the boundaries of discrete farming areas and become concentrated in populated lands or water sources centers far distant. Ecologists call this phenomenon non-point source (NPS) pollution, meaning that the cumulative effect becomes

---

A CASE OF A BLUE BABY BOY

A one-month-old infant boy was rushed to the emergency department of a rural hospital. His parents stated that recently he had become blue-colored and began suffering breathing difficulties, struggling to take in air. Despite being placed on a full mask in the ER through which he breathed 100 percent oxygen, the baby had cool and dusky skin, with low measured oxygen saturations. Recognizing that the infant was profoundly ill, but not knowing the precise cause, the treating team warmed him, administered a dose of antibiotics, and arranged for helicopter transport to a regional medical center.

The accepting team at the medical center did a full laboratory evaluation that included arterial blood gas analysis. The blood appeared chocolate-colored, reflecting the poisoned red blood cells. The methemoglobin level measured 92 percent, with a normal level being approximately 1 percent. The infant was quickly given the antidote, methylene blue, along with supportive care, and therefore enjoyed a complete recovery. Ensuing investigation concluded that the baby had been unintentionally poisoned because his formula was mixed with water drawn from an old well located on the family farm. The water contained dangerously high concentrations of nitrates, attributed to agricultural runoff from recent heavy rains being absorbed through the surface soil and leaking through cracks in deteriorating walls of the well, thereby contaminating the water.

concentrated contamination from several different sources. NPS often describes the effects on watershed areas that collect agricultural pollutants in wetlands or other bodies of water.

High concentrations of nitrates (a key ingredient in fertilizer) in drinking water can cause a blood disorder known as *methemoglobinemia*. Nitrates poison red blood cells by forming methemoglobin, a form of hemoglobin that does not bind and carry oxygen. It effectively can be likened to suffocation. People with severe methemoglobinemia take on a bluish coloration to their skin. Babies are most susceptible to the adverse affects of elevated methemoglobin, typified by the "blue-baby syndrome."

During heavy precipitation, farm soil from a large watershed area can overwhelm nearby bodies of water with heavy loads of fertilizer and sediment. Concentrated amounts of fertilizer act like a gigantic growth hormone bolus to aquatic life, leading to harmful algal blooms (HABs) and a toxic risk to humans.[56] Cyanobacteria, sometimes misnamed as blue-green algae, produce toxins that poison drinking water and can cause gastroenteritis, skin irritation, allergic rashes, and liver damage.[57] *Pfiesteria piscicida* is a single-celled alga usually found in estuaries; it has been linked to HABs that cause headache, confusion, rash, and eye irritation.[58]

Sewage and Environmental Contamination

The final ubiquitous toxin is human excrement, which is a constant in our existence. One of the bedrock principles of public health is to be able to separate what and where we eat and drink from where we urinate and defecate. A constant threat of food contamination exists in places where excrement is concentrated, such as sewage treatment areas. Improperly contained sewage leaches, leaks, spills, or flows into nearby neighborhoods or farmland. Accidents happen in part because forces of nature distribute excrement from containments that cannot withstand the magnitude of an extreme weather assault.

Most municipal systems, at least in wealthier nations, are designed to daily handle megatons of human waste. However, they barely have kept pace with the tremendous global urban expansion of the past few decades. Many urban sewage networks now operate at maximum capacity. They malfunction, sometimes to the point of overflowing, during what appear to be innocuous rainstorms. This limitation of infrastructure, originally designed and built for populations far smaller than what they now serve, is compounded by the fact that most urban areas do not absorb rainwater uniformly or consistently. With decreasing amounts of arable soil, the land's absorption capacity continues to diminish. The result too often is a situation whereby a summer rain shower in a city stresses a sewage system to the point of dangerous runoff of untreated excrement and toxic material, contaminating local water sources and public congregating areas. The fact that extreme downpours are now increasing around the globe makes this a greater concern.

The public health implications of sewage overflows are tremendous. The Environmental Protection Agency estimates that each year, between 1.8 and 3.5 million Americans are estimated to become ill from recreational contact with sewage from sanitation sewer overflows.[59] Common illnesses caused by swimming in, bathing in, or drinking untreated or partially treated sewage include skin infections and rashes, and gastroenteritis. Chronic exposures are linked to cancer, heart disease, and arthritis.[60]

CONTAINMENT

Effective containment is the key to living in a world full of toxic chemicals, but extreme weather and flooding are formidable challenges. Many exposures occur when containers, whether stationary or portable, malfunction because of improper transfers or leaks. Even when containers meet regulatory specifications, an outside disaster may strike and expose them to stressors well beyond their structural and functional capabilities. Consequences to the environment and human health can be devastating.

Louisiana is a hub of crude oil refining in America. When Hurricane Katrina struck in 2005, extreme winds and storm surge disrupted a 250,000-barrel storage tank, releasing approximately 1.05 million gallons of crude oil into the adjacent residential neighborhoods and canals. The "Murphy Oil Spill" gushed and spread to involve approximately seventeen hundred homes with significant deposits of heavy metals and organic hydrocarbons that penetrated into the soil.[61] Following the floods in central Europe in 2002, the persistent organic pollutants polychlorinated dibenzo-p-dioxins and dibenzofurans were found to have leaked from containers into the surrounding soil. The toxic substances subsequently were discovered to have bioaccumulated within the food chain, such as in regional cow's milk.[62]

Our global human population is more than seven billion. This is projected to increase by more than one billion people within the next fifteen years, reaching 8.5 billion in 2030, then 9.7 billion in 2050 and 11.2 billion by 2100. That's a nearly unfathomable number of people to feed! When we consider that the impact of climate change and other man-made alterations to our environment easily could neutralize any future gains in food productivity, in all possibility the consequences of undernutrition will top the list of the most widespread health effects of climate change.

· *10* ·

Allergens

$\mathcal{D}$uring "pollen season," many of your friends and family will be sniffling, coughing, and wheezing, with runny noses and itchy watering eyes. Chances are they're being dominated by airborne allergens. For many of these people, spring is not a season of birth and renewal but, rather, a time of dread. Allergies, seasonal and otherwise, are among the most common medical conditions worldwide. In the United States alone, an estimated fifty million persons suffer from bronchitis, asthma, or one of many other allergic conditions, all courtesy of the dynamic duo of the airborne allergens pollen and mold.[1] In the bodies of sensitive sufferers of allergies, a chain reaction of debilitating symptoms originates when histamine and other chemical mediators activate an inflammatory cascade that swiftly swells the mucous membranes inside the nose, inflames the capillaries of the skin, and constricts the smallest airways of the lungs. These symptoms can persist and progress to aggravate breathing disorders, such as asthma and chronic obstructive pulmonary disease.

Allergens have a significant impact on health worldwide. Therefore, any phenomenon that increases their presence or propensity to provoke human allergic reactions is a cause for concern. A warmer world with weather extremes and atmospheric turbulence that more widely distributes allergens likely will have more people laid low by seasonal allergies. A rise in atmospheric carbon dioxide concentration will amplify production, allergenic content, and dispersion of aeroallergens, and promote a thriving habitat for organisms that produce them (e.g., weeds, grasses, trees, and fungus). It also has been demonstrated that air pollutants can act alone or in synergy with aeroallergens to exacerbate allergic diseases.[2]

Aeroallergens generally are attributed as the source of exposure for three of the most common allergic ailments: allergic rhinitis ("hay fever"), asthma, and atopic dermatitis (eczema). The incidence of all of these conditions is on the rise.[3] Simultaneously, worldwide pollen production has spiked, increases in fungal growth and spore release have increased, and human sensitivity to allergens has increased.[4] Approximately 15 to 20 percent of children worldwide are afflicted with atopic dermatitis, and current trends support that this percentage is increasing.[5]

With regard to the plentitude and health of plants, our planet is a very complex system. Therefore, any interpretation of the interaction between climate change, plant life, and human health effects is more than a simple understanding that elevated temperatures and more carbon dioxide will yield a world full of more plants, and therefore a proportional increase in pollen and allergies.[6] Similarly, it would be fallacious to assume that climate change provides a silver lining of our entering into a "greenhouse era" of lush increases in both diversity and amount of flora. However, it is reasonable to predict that if they become significant, detrimental effects of increased allergens in the air we breathe will become more apparent. So, let's consider how the offending agents comprised of pollen, molds, and botanical skin irritants will react to evolving environmental conditions.

POLLEN

Continued existence of a species requires reproduction. Based on the amount of pollen in the world, plants are very good at it. Pollen is the genetic material that plants release into the air (or as "payloads" onto pollinators, such as bees) in hopes of fertilizing other members of their species. They most often employ a "shotgun" technique, spewing sometimes immense clouds of fine reproductive dust that can be so dense that they blanket everything in the surrounding area. For example, a single ragweed plant can generate millions of pollen grains in a single day. Pollen is nearly always light, dry, and custom-made for wind transport, so that it may travel hundreds of miles from the plant of origin. By examining changes in pollen production and dissemination caused by a changing climate, palynologists (botanists who study pollen and spores) have found that plants have expanded their growing seasons and prolonged the duration of releasing allergenic pollen.[7] They've identified several examples where regional increases in precipitation and temperature have significantly influenced pollen production. For instance, maximum pollen counts were higher during a warmer, wetter El Niño year than in the year before or after it in New England; oak pollen counts in the San Francisco Bay Area showed strong correlations with total rainfall from the previous year; and increased pollen counts of cypress pollen in Oklahoma displayed significant positive correlations with elevated temperatures and negative correlations with precipitation.[8,9,10] Another landmark study declared that the ragweed pollen seasons measured between 1995 and 2011 have now lengthened throughout North America, by as much as twenty-seven days in the most northern latitudes.[11]

Extreme weather events portend interaction between climate, aeroallergens, and allergies. Thunderstorms are known to be a trigger for higher rates of asthma attacks—both in asthmatics and in those who previously have had only mild allergic conditions.[12] Such violent storms and their associated winds and heavy rains foster optimum conditions for pollen grains to rupture, consequently releasing small particles that induce shortness of breath from small airways irritation.

What Is a Pollen Count?

The pollen count is a favorite metric of television weather commentators during spring and summer months. It's a measure of pollen in the air and thus correlates with symptoms of seasonal allergy disorders. It is measured as grains of pollen per square meter of air over a twenty-four-hour period. Pollen counts tend to be highest on warm, dry, windy days and lowest on cool, damp days. The numbers are usually highest in the morning hours. What does the future hold? A recent presentation at the American College of Allergy, Asthma and Immunology suggests that for reasons in part attributed to climate change, pollen counts will more than double by 2040.[13]

A VERY CLOSE CALL

Amanda's asthma could be triggered by lots of things—warm weather, cold weather, pollen, excitement—even being frightened at the movies. She had her first near-death experience in junior high school, when an attack of difficult breathing seemingly came out of nowhere, although afterward she recalled that she might have been stung by a bee. The truth was that everyone's seasonal allergies were worse, because species of grass and trees were proliferating in the neighborhood that hadn't been there decades previously. It was simple arithmetic: the more allergens, the greater the number of persons with allergies; and the more people with allergies, the more opportunity for a serious reaction.

This was "extrinsic" asthma, caused by external factors that weren't strictly related to genes and the family tree. Combined with Amanda's inherited propensity to intrinsic asthma, it could be a one-two knockout punch. She was on a boatload of medications to suppress the cascades of immunity and inflammation that caused her airways to become swollen and constrict, but once a severe reaction got going, it could flare beyond control in minutes and cause her breathing to shut down. It became a vicious cycle of decreased oxygen, effectively smothering her, and a panic reaction. Breathing faster or with more effort wouldn't help. Anyone who has suffered a severe asthma attack knows that darkness comes from all directions, and sometimes the only light is a tube in the trachea and being hooked to a breathing machine.

She strolled into the locker room with a slight wheeze, which wasn't unusual. She often walked a tightrope in the late spring, when blossoms were beautiful and the new-mown grass smelled fresh but heralded allergies. A couple of pills and a puff or two on her inhaler normally would suffice to allow her to be with her classmates and take a few easy laps around the track. Today was like any other day, except it was unseasonably warm, the infield grass had just been cut, and the girls left footprints in the yellow dust that blew off the pine trees. She was tired, which didn't help, and had the remnants of a lingering upper respiratory tract infection. Then, she encountered an extrinsic barrage. Had an epidemiologist been in her county doing surveillance, she would have found twice the usual number of asthma attack victims who came to the ER that week. Had she been able to determine the confounding factors, they would have been the plants and the heat, each in and of itself powerful, and in combination synergistic in a very bad way.

The first lap was manageable, but by the time she was once again halfway around the track, it hit her. Something was dreadfully wrong, so she bent over, hands on knees, to try to catch her breath. The harder she tried, the worse it got. Her classmates ran a full lap until they came upon her again, crawling and struggling to stay calm. She vomited, her lips turned blue, and she began to choke on her saliva. Just before she became unconscious, she managed to whisper, "EpiPen," and someone had the presence of mind to race to the principal's office, where they kept emergency medical supplies. Her gym teacher, who had been taught how to use the device, held it tight against Amanda's thigh and triggered the needle to enter the muscle and discharge its contents. The medication was effective in the nick of time. After she could have lost her life, Amanda recovered to jog another day.

MOLD

Mold makes up the other half of the dynamic duo (with pollen) of environmental allergens. The kingdom *Fungi* entails life forms distinctly different from plants and bacteria, and includes mushrooms, molds, yeasts, rusts, puffballs, truffles, morels, and smuts. They tend to grow in moist, warm, and often shady areas but are found nearly everywhere, including indoor environments, such as shower stalls or areas that have water leakage. The reproductive units of fungi are called *spores*. Like pollen grains, spores are very small, abundant, and highly allergenic. Inhalation is the most common type of exposure to mold. Because many airborne fungal spores are tiny, they are inhaled into the deepest parts of the airway and lungs. Allergic reactions to mold spores are typified by symptoms similar to those associated with pollen allergies: eye irritation (conjunctivitis), runny and itching nose with nasal congestion (allergic rhinitis), inflammation within the sinuses (sinusitis), and asthma.

Where humidity and precipitation are found, particularly in combination, so is mold. Warming environmental temperatures, along with increased carbon dioxide and extreme weather events, will be considerable drivers for proliferation of allergenic molds. The impacts on human health can be substantial. Molds are diversely effective pathogens; they not only cause allergies but can also undermine human health in nonallergenic ways. In damp conditions, mold can rot grain stores rapidly, which effectively destroys tremendous amounts of grains, such as wheat, corn, and barley. This undermines food security in a devastating manner. Major risks for mold exposure arise during periods of sustained humidity and in flood or post-hurricane/typhoon/monsoon environments. Massive water deposition into residential areas by heavy rains and standing pools of water can lead to dangerous mold levels, particularly when homes have had water damage. Sediment deposition in flooded spaces that retain water and moisture creates a protected, wet environment that is a haven for fungal growth, which translates into a toxic milieu for humans. The threat to a person's lungs is greatest during cleanup or repairs, when mold deposits are disturbed and then can easily enter the air we breathe and, therefore, invade the smallest airways within our lungs.

Most recent data on mold exposure attributed to an extreme weather event come from New Orleans in the aftermath of Hurricane Katrina. A significant percentage of city buildings were underwater for weeks, resulting in a proliferation of mold. According to the Centers for Disease Control, nearly half of the homes in the New Orleans area showed visible mold growth, 17 percent had heavy mold coverage, and allergy-causing molds were found in both indoor and outdoor environments.[14] Outdoor spore counts remained abnormally high for months after the hurricane, with higher counts recorded from persistent mold growth inside flood-damaged homes. The medical clinical aftermath was profound. Local hospitals reported increased numbers of patients complaining of allergies and cold symptoms. A spike in complaints of chronic cough associated with difficulty breathing was nicknamed "Katrina cough" or "shelter cough."[15] The overall rate of childhood asthma increased in the New Orleans area after Hurricane Katrina.[16]

Predictions are that because of climate change, we'll encounter many more extreme storms and urban floods in the coming century. If that is the case, effects upon human health will be very likely. The lessons of Hurricane Katrina comprise an example of what we can expect from a warmer, moldier world.

Heat and dry weather promote other fungi. Coccidioidomycosis ("Valley fever") is a widespread fungal disease caused by *Coccidioides immitis and C. posadasii*, which can be found in the desert soils of Arizona, southern California, New Mexico, Texas, and parts of northern Mexico. Arthroconidia are a certain type of spores created by fungi. These spores are activity-dormant during the long, dry spells that are typical of the American Southwest. Then comes a brief rainy season, which triggers a massive release of these spores. They become airborne, so can be inhaled and cause an array of deleterious effects. Most of the time, the microbial ecosystem, including bacteria and other organisms, keeps derivative pathogens relatively in balance and therefore from unduly proliferating. However, as the climate changes to yield warmer seasons and unrelenting droughts, natural biological checks and balances alter in such a way that fungi can form greater numbers of spores, many of which survive the harsh environment and travel far away with the wind. The number of cases of coccidioidomycosis has increased over the past two decades.[17] Causality is thought to be multifactorial, but the crushing California drought (according to one source, the worst drought conditions in the past twelve hundred years) is at the top of the list of causes.[18] Serious complications include pneumonia, inflammatory lung disease, and "disseminated disease," a severe full-body infection in which the spreading fungus can cause skin abscesses, joint pains, vital organ inflammation, meningitis, and death.

AIR POLLUTION AND AEROALLERGENS

Our understanding of the interactions between air pollution and aeroallergens suggests that they synergistically contribute to disease potency. One reason may be that the air pollutant ozone, particulate matter (PM), and sulfur dioxide separately and together cause airway inflammation that facilitates dispersion of pollen allergens deep into the airways.[19] Air pollutants have other unfortunate characteristics. They can increase release of antigens (substances that lead to allergic responses) in pollen grains and also can absorb pollen grains and act to prolong their retention in the body after they have been inhaled, effectively creating a sustained-release antigen.[20] As both ground-level ozone and aeroallergens concentrations are predicted to increase with rising average temperatures, the interplay between these two airway irritants will be an important determinant of our future respiratory health.

Inflammation of the Skin (Dermatitis)

The suffix "itis" connotes inflammation. Skin rashes can be a manifestation of inflammation, or they may be caused by direct injury (e.g., a burn) or a bite (e.g., from a mosquito). Many plant species have developed defensive chemicals that are harmful to humans. By far the most common to us are those that cause "contact dermatitis." Examples are the rashes of poison ivy, oak, and sumac. It is estimated that two of every three persons are sensitive to urushiol, the principal allergen produced in these plants.[21] Rising temperatures and elevated carbon dioxide levels have been demonstrated to increase both the geographic rate of spread and above-ground biomass of the plants, as well as the concentration of urushiol within the plants.[22] In this regard, climate change essentially is allowing development of much more potent and harmful strains of poison ivy, oak, and sumac.[23]

Not all plants will thrive equally in a warmer world with a more carbon-rich atmosphere. Biodiversity is predicted to suffer because species will be eliminated. For instance, ragweed and poison oak/ivy/sumac may thrive and proliferate, but other species will struggle to adapt. The resulting scenario is that we will have a more unpredictable natural world, dominated by fewer—but aggressive—plant species characterized by longer pollen seasons, heavier pollen loads, and greater allergenicity to humans.

In a moment of stress, you may say to someone, "Just take a deep breath." This follows the historical notion that fresh air is a salve to most of life's problems. If our environment changes, we may need to say, "Just try to hold your breath."

• *11* •

Harmful Algal Blooms

*W*e've discussed the marine ecosystem in the context of water security. Equally important are natural inhabitants of the water. These are not the diarrheal germs and other pathogens that use water very effectively as a route to invade our gastrointestinal tracts but, rather, the copious and diverse array of algae that are an integral part of the aquatic ecosystem. The term "algae" refers mostly to water-based organisms that use sunlight to produce carbon energy (photosynthesis) and that lack the distinct tissue characteristics of plants. Algae range from microscopic one-celled life forms (e.g., diatoms of the genus *Chlorella*) to giant kelp (*Macrocystis pyrifera*) that can grow to fifty meters in length. Without algae, the upward food chain of marine life would grind to a halt, because algae provide the foundation upon which this chain is built. Algae use sunlight to generate other sources of energy and essentially become an originator of essential carbohydrates and proteins for predatory higher order organisms (e.g., shellfish, fish, waterfowl, and ultimately apex predators, such as carnivorous mammals). Their existence is essential to support life on Earth. However, when external forces, such as a change in nutrients, temperature, or currents, affect their environment, certain species of algae will respond with a positive-feedback loop of unchecked growth and rapidly can overwhelm the aquatic ecosystem. As algae grow in numbers, so too does proliferation of their chemical products, certain of these toxic to humans. Thus, we witness creation of the harmful algal bloom (HAB).

CLIMATE CHANGE AND HARMFUL ALGAL BLOOMS:
WARMING WATERS AND EXTREME WEATHER

The evidence is compelling that algal blooms are occurring more frequently and across a wider area of the planet than at any other time in recorded history.[1] The cause is a combination of increased agricultural nutrient use, climate change–driven extreme weather, water temperature rise, and elevated atmospheric carbon dioxide. A warmer planet and resulting increase in ocean temperatures will prolong the calendar time period during which seasonal blooms occur, expand algae distributions to more temperate parts of the world, and provide competitive edges for the growth of specific algae known to be toxic to humans.

For example, the algae phylum *cyanobacteria* (also known by the misnomer "blue-green algae") has shown particular adaptability to such changing conditions. Recent data clarify that water temperatures in excess of 25°C (77°F) allow optimal growth rates for cyanobacteria, causing them locally to dominate other species and more readily blossom into (harmful) algal blooms.[2] *Gambierdiscus toxicus*, the dinoflagellate algae responsible for ciguatera fish poisoning, has been implicated in increases in human poisonings where sea surfaces are warming regionally.[3] Cyanobacteria, *G. toxicus*, and *Alexandrium catanella* (the algae linked to paralytic shellfish poisoning) have all been predicted to expand their range into temperate regions as water temperatures rise, increasing the risk to human health.[4]

Warmer temperatures will add energy to our weather systems. The resulting hydrologic extremes likewise will favor more HABs. Intense and longer rainstorms may further enhance cyanobacteria growth through greater nutrient loads from agricultural runoff. Over the past twenty years, alternating heavy rain and drought cycles have increased worldwide.[5] These cycles serve to create a "bolus effect," as built-up nutrients are washed downstream from a sudden deluge of intense precipitation that causes heavy water flow, including floods over parched, desiccated lands.[6] Droughts and rising sea levels are expected to continue to increase the salt content of water in many parts of the world. *Anabaena, Microcystis,* and *Nodularia* are all salt-tolerant cyanobacteria that are more suited than other freshwater species to increased salinity. Human users of waters that contain high concentrations of their toxins are at risk of adverse health effects, such as liver and nervous system toxicity, commonly caused by cyanobacteria HABs. As droughts become more common, freshwater is at risk for becoming saltier due to less flow and decreased water volume. This provides an ecological niche for potentially toxic marine algae to invade freshwater ecosystems. We already have witnessed this in the southwestern and south central United States, where marine algae have infiltrated non-native areas.[7]

CLIMATE CHANGE AND HABS:
ELEVATED CARBON DIOXIDE

For life on Earth that depends on photosynthesis, carbon dioxide is essential. Carbon dioxide is to these life forms what oxygen is to humans. Since the year 1850, carbon dioxide in the atmosphere has increased from 270 parts per million (ppm) to more than 400 ppm. The vast majority of this increased carbon dioxide is sequestered in the world's oceans, leading to a decrease in the pH in the ocean. A lower pH means a more acidic environment relative to a higher pH. These changes in acidity favor the growth of certain species (e.g., *A. catenella* and *Pseudo-nitzschia*) and their toxic products.[8] Ocean acidification has been linked to disruption of the chemistry of calcium carbonate, the essential compound for calcification of certain organisms, such as coral.[9] Over the past decade, biodiversity of coral reefs has decreased as the reefs have declined, which is attributable partially to warming water temperatures and ocean acidification. Widespread "bleaching" of tropical reefs is a visible reminder of natural destruction that might accelerate, with potentially devastating human health consequences from disruption of food security, ecosystem services, and biodiversity. Furthermore, a more acidic ocean environment will favor organisms, such as *dinoflagellates*, that are the phytoplankton group comprising the most harmful algae to humans.[10]

WHAT IS *HARMFUL* ABOUT AN ALGAL BLOOM?

FOOD FOR THOUGHT

Walking on the beach in Oregon in September was soothing for John's soul. It was unseasonably cool, brisk enough to see your breath, but not to paint a frost on the leaves. The water was still warm from summer, and the seabirds frolicked as they darted into mats of seaweed to pluck baitfish from below the ocean surface. Deeply black-blue mussels clustered on the rocks, where they were exposed by the waning tide. He couldn't resist prying free a few, placing them in a small bucket he carried mostly to collect shells of dead crustaceans and mollusks, stray pieces of driftwood, and colorful stones. He'd cook the mussels in a stew with wine, lemongrass, and a dash of curry. He didn't know that a few days earlier, the water wasn't clear, but discolored by a brownish-red "tide" produced by plankton in the ocean. The mussels he harvested had dined on these microorganisms and then concentrated poison in their digestive organs and muscle tissues.

At home, the fragrant improvised stew had John salivating. He wolfed down a large bowl of gourmet goodness. As he explained to the medical team that saved his life after he finally woke up, minutes after eating the contaminated shellfish, he felt "weird," as if he were under the influence of a sorcerer. In rapid succession, he felt numbness and tingling inside his mouth, and of his tongue and gums. He soon became lightheaded, weak, and incoherent, and began to drool, unable to swallow his saliva. Trying to dial his phone for help, he was completely without muscular coordination. His vision blurred, his belly hurt, he had a splitting headache, and his heart was racing. Just before he suffered total paralysis, he managed to stumble outside, where a neighbor saw him fall and called 911. The medics arrived in the nick of time. They didn't know the cause, but they knew how to help a man who couldn't breathe and saved his life by placing a tube down his throat into his trachea. Five minutes later, and he would have suffered catastrophic brain damage from lack of oxygen.

We've outlined the case for a future where the changing climate will cause more frequent and widespread algal blooms. What is the health risk? To date, HABs have had a relatively small impact on humans, compared to infectious diseases such as malaria and tuberculosis. As more events occur, we'll refine our understanding. Here is what we know already.

Cyanobacteria

Cyanobacteria are present in all freshwater ecosystems and are predicted to play a prominent role in climate change–driven HABs. They produce *cyanotoxins*, a group of structurally diverse compounds that are highly poisonous to the human nervous system, liver, kidneys, and skin.[11,12] Human exposure can occur through drinking affected water or by direct aquatic exposure from fishing, swimming, water skiing, and so forth. The toxic effects of cyanobacteria are diverse and formidable. One type of cyanotoxin acts upon the nervous system, which in turn affects skeletal muscles, causing staggers, muscle twitching (fasciculation), and convulsions (seizures). HABs can accumulate concentrations of toxin sufficiently potent to fatally paralyze small mammals. The toxin works in much the same way as a powerful drug that physicians use to paralyze patients intentionally. Other cyanotoxins directly cause liver cell death. The toxins have multiple effects upon humans, and symptoms run the gamut from gastrointestinal distress to difficulty breathing to itchy skin rashes. According to the United States Geological Survey, cyanotoxins have caused human illness in more than fifty countries, including thirty-six states in the United States.[13] A report from Brazil in 1996 cited more than fifty deaths from liver failure when local cyanotoxin-contaminated water was unknowingly used for kidney dialysis.[14]

Another HAB-related cyanotoxin syndrome has been named *Haff disease*. This is characterized by severe muscle breakdown in people who ingest an affected species of the bottom-feeder buffalo fish (genus *Ictiobus*).[15] Symptoms include diffuse muscle pain, tenderness, rigidity, and the condition of rhabdomyolysis, where muscle tissue breaks down into its protein constituents. When this happens, kidney failure can occur, because the copious protein load in the bloodstream acts to "clog" the filtering kidneys.

Red Tides

Colored "tides" are events that literally can take away your breath. Red tides derive from proliferation of dinoflagellates in the genera of *Alexandrium* and *Karenia*, whose blooms are of a reddish or brown color. According to the National Oceanic and Atmospheric Administration, red tides are present in every coastal state, and their incidence is increasing.[16] *Karenia brevis* produces *brevetoxin*, a type of nerve toxin that can kill fish and other marine animals. Humans who consume shellfish that have bioaccumulated these toxins can suffer severe symptoms of painful tingling in the fingers and toes, or gastrointestinal symptoms, such as vomiting and diarrhea. Direct exposure to water is not the only form of transmission. Brevetoxins can become airborne in an aerosolized form during HABs and contaminate the surrounding air and anywhere downwind, up to one mile inland.[17] They are known to induce swelling and spasm of the human respiratory tract (bronchoconstriction), causing an asthma-like sickness characterized by wheezing and shortness of breath. People with preexisting asthma are most at risk and may experience asthma attacks for days after the initial exposure. Red tide events have been correlated with spikes in emergency department visits for asthma, bronchitis, and pneumonia.[18]

The direct health effects of red tides are known and quantifiable, but less so the anxiety they produce. It is observed that they can impact mental health significantly and may act as a threat multiplier for civil discord and food insecurity. In 2016, a red tide HAB off the coast of southern Chile, thought to be secondary to warmer waters because of an exacerbation of an El Niño weather oscillation, became one of the worst on record and contributed to massive seafood die-offs.[19] The marine mortality caused panic among the public, with fishermen taking to the streets, blocking roads, and publicly denouncing the government. A state of disaster was declared, and the government allocated 100,000 Chilean pesos (approximately US$150) to each impacted family deprived of their livelihood from the collapse of a major food industry.

Shellfish Poisonings

Dinoflagellates and other species of marine microalgae known as *diatoms* (characterized by dual-chambered cell walls) produce some of the most powerful natural toxins known to science. These *phycotoxins* are incredibly hearty and able to withstand cooking, boiling, freezing, and drying. Syndromes caused by these toxins include paralytic shellfish poisoning (PSP), neurotoxic shellfish poisoning (NSP), amnesic shellfish poisoning (ASP), and diarrhetic shellfish poisoning (DSP).

Ciguatera foodborne illness ("ciguatera") is a dinoflagellate-caused poisoning that is one of the most common nonbacterial causes of seafood poisonings in the United States. It results from eating tropical and subtropical reef fish that have bioaccumulated ciguatoxins, which are produced by several marine microorganisms, most notably the dinoflagellate *Gambierdiscus toxicus*. *G. toxicus* adheres to dead coral and marine algae consumed by smaller herbivorous fish. These fish are preyed upon by larger reef fish, which are in turn eaten by larger fish, bioaccumulating the toxin up the food chain. The link between human gastronomy and potential exposure is well known. The list of potential offenders is found on many neighborhood seafood restaurant menus and includes amberjack, snapper, tuna, mahimahi, seabass, and grouper.

Poisoning occurs in a dose-dependent manner, meaning that the more contaminated fish one consumes, the more symptomatic one will become. The unsuspecting victim will not notice any change in the taste or smell of the fish, which can appear entirely normal. After ciguatera toxin(s) is consumed, physical symptoms can occur within minutes. These include nausea, vomiting, diarrhea, headache, and muscle pain. Severe cases can include neurologic symptoms of numbness, loss of balance, and hallucinations. Ciguatera poisoning is infamous for its unusual physiologic symptoms, such as *cold allodynia*, which is the sensation of burning on one's skin from contact with a cold object. Other symptoms include complaints of the sensation of loose or painful teeth. Ciguatera can cause disability that lasts weeks to years. Death rates of 0.1 to 12 percent have been reported, with respiratory paralysis being the most common fatal condition.[20]

Ciguatera has no cure, and although most people recover, they are susceptible to redeveloping their symptoms in the future. Commonplace food and activity that can trigger a recurrence include eating nuts, alcohol, fish, or eggs, and exercise. Rarely, ciguatera poisoning can pass between people through sexual transmission. A case documented a mother who noticed a change in behavior of her three-year-old child, along with changed sensitivity to hot and cold temperatures, following breast feeding.[21]

A toxin similar to that found in ciguatera has been implicated in diarrhetic shellfish poisoning (DSP), which has been reported in Japan, Europe, and Chile. Onset is fast and violent, with vomiting, diarrhea, and chills that may last up to three days. Although its acute symptoms are more mild than those of ciguatera, some evidence suggests that the toxins are tumor-promoting agents.[22] Skin itching has been reported to persist for weeks after poisoning and worsen with activities that increase skin temperature (exercise, consumption of alcohol).[23]

Such illnesses can be so disruptive and frightening that many people will avoid seafood altogether after recovering from a poisoning. One thing is certain—they certainly will never forget the experience, unless the toxin erodes their memory. Phytoplanktons of the genus *Pseudo-nitzschia* can cause a syndrome named amnesic shellfish poisoning (ASP). These algae, which are diatoms, contain domoic acid, a neurotoxin poisonous to humans. Domoic acid bioaccumulates in shellfish; when consumed, it is absorbed and subsequently enters into the human brain, affecting nervous system signal transmission. It was first documented when 150 people became ill after eating mussels on Prince Edward Island, Canada. Four people died, and many others remained afflicted with persistent memory problems.[24] Researchers later ascertained that domoic acid has a dangerous affinity for neuroreceptors in the brain, particularly in the memory centers of the hippocampus and thalamic region.[25] This particular seafood poisoning begins with gastrointestinal distress, followed as soon as forty-eight hours later by memory impairment. The most common deficit associated with ASP is difficulty in processing new memories (although some severe cases also showed loss of previous memories). Rare symptoms are seizures, partial paralysis, persistently blurred vision, and coma.[26,27]

Rounding out the seafood blues is another disturbing syndrome called paralytic shellfish poisoning (PSP), caused by the algal poison saxitoxin. PSP can be a fatal condition, especially in children or the immunocompromised. Within ten minutes of ingesting the neurotoxin, humans can suffer vomiting, abdominal pain, and diarrhea, along with tingling and burning sensations in the face, body, and limbs. Symptoms may progress to shortness of breath, slurred speech, and confusion; trunk and extremity weakness over a couple of hours; and paralysis shortly thereafter. If paralysis extends to the diaphragm and chest wall muscles, breathing will become all but impossible. Death rates of up to 50 percent in children previously have been reported, dropping in recent years due to improvements in supportive medical care.[28] Toxicologists consider saxitoxin one of the most powerful nonprotein poisons.

BLUE-GREEN ALGAE AND LOU GEHRIG'S DISEASE

An increasing suspicion has emerged about a linkage between amyotrophic lateral sclerosis (ALS), known as Lou Gehrig's disease, and exposure to high levels of cyanobacteria. ALS is a progressive neurodegenerative disease that affects the brain and spinal cord. Often it is fatal. Researchers at Dartmouth-Hitchcock

Medical Center identified an ALS cluster around Mascoma Lake, New Hampshire. They hypothesize that this disease cluster (with a local population showing disease at twenty-five times the national prevalence) and similar ones around lakes in northern New England stem from long-term exposure to a neurotoxin from cyanobacterial blooms.[29]

Research has focused on the amino acid beta-methylamino-L-alanine, which can be produced by cyanobacteria.[30] This amino acid has been found in brain biopsies of ALS patients and other people with neurodegenerative diseases. Similarly, veterans of the Gulf War (1990–1991) have shown a higher incidence of ALS than have other military personnel.[31] One biologically plausible explanation points to the cyanobacterial ground crusts and mats in the region. Researchers have hypothesized that increased military vehicular activity disturbed the ground and created increased airborne exposure to cyanobacteria.[32]

Harmful algal blooms are one manifestation of a natural phenomenon made worse by global warming, because of the frequency and extent of the blooms. We will not eliminate algae from our environment, nor should we even contemplate that as a solution. They are an essential component of perhaps the most important food chain on Earth. Rather, it is the mass episodic coastal proliferation of these microscopic life forms that is unnecessary and unwanted. The root causes are the temperatures of the sea and air. To the extent that these are under our control, we should seek to best mitigate them as risk factors.

Part III

BIODIVERSITY IN DESCENT

· *12* ·

Nature's Pharmacopeia

Consider this: Without the medicinal gifts that a biodiverse world offers, our heads would throb and burn. Our hearts would deteriorate and fail prematurely. We might die from the most common infections. If you take a medicine, chances are that the drug was derived from or modeled after a chemical from a plant or fungus. Between 1998 and 2002, 116 of the 158 latest small-molecule drugs approved by the U.S. Food and Drug Administration originated directly or conceptually from nature.[1] This small number is just a tiny fraction of what is available and yet undiscovered in the vast warehouse of nature's pharmaceuticals. The biochemical repository in our gardens, forests, and oceans has improved and prolonged our lives.

Biologists, ethnobotanists, and biochemists continue to identify potentially beneficial chemical compounds from nature. Meanwhile, Earth loses species at up to one thousand times above background extinction rates when compared to historical records, according to Dr. Eric Chivian, director of the Biodiversity and Human Health Program in the T. H. Chan School of Public Health at Harvard University.[2] Dr. Gro Brundtland, former director-general of the World Health Organization, proclaimed, "The library of life is burning, and we don't even know the titles of the books."[3]

Over a time span of more than 3.5 billion years, the laboratory of nature has evolved a wondrous panoply of chemical compounds, such as enzymes and hormones, most of them presumably to defend certain species from external (e.g., thermal conditions, predators) and internal (e.g., infections) threats. The compounds presumably were derived to confer an evolutionary advantage, be it growth, longevity, or resistance to a specific harmful condition. In many cases, the activity and benefits of these compounds extended beyond the originating species to other plants and animals. Thus, poison from a tree frog might have benefit as a medicinal paralytic agent in humans. Whether an elm tree is fighting parasitic fungi or a cone snail is capturing its prey, opportunity exists for human benefit. If we eliminate the original species, all of this potential is lost. The South American cinchona tree and the Chinese sweet wormwood plant effectively treat and cure malaria. With only rudimentary refinement, quinine was used routinely hundreds of years ago to combat malaria. Poppies produce analgesics to treat pain that once was unbearable and debilitating. Some of the alkaloids from the Madagascar periwinkle (*Catharanthus rosea*) yielded essential compounds used in cancer chemotherapy. If we lose the plants we have yet to discover and analyze, we will not realize their benefits.

The Greek physician Hippocrates wrote of an elixir concocted from leaves of the willow tree. He knew this to be an effective analgesic and described its administration to lessen the pain of childbirth. Over more than two millennia, many cultures used this remedy as an herbal tea by brewing the bark to treat symptoms of infection, injury, and arthritis, but it wasn't until the mid-eighteenth century that its mechanism was revealed.

Edmund Stone, a British Episcopal priest, painstakingly monitored use of the bark of white willow (*Salix alba vulgaris*) to treat fevers. As the story goes, Mr. Stone had an interest in the ancient "Doctrine of Signatures," which espoused that *morphological plant signatures*, or plants that resemble part of human anatomy, can be used to treat illness or injury specific to that anatomical site.[4] Such belief influenced early characterization of the natural world: *bloodroot*, *liverwort*, *lungwort*, *snakeroot* (i.e., a cure for snakebites), *toothwort*, and *wormwood* (to treat gastrointestinal "worm" parasites). The inquisitive Stone obtained a small piece of willow tree bark and was impressed by its extremely bitter taste. Curious because of its known therapeutic potential, he began to experiment. The Victorian age did not have today's strict ethical guidelines for medicinal experimentation, so Stone moved quickly from "bench" research to human studies. He soon collected and dried more than a pound of willow tree bark, creating a powder that he then fed to scores of people. The effect was a "powerful astringent and very efficacious in curing agues and other disorders," including reducing fevers.[5] The precursor to aspirin was born.

In the nineteenth century at Friedrich Bayer's lab in Germany, Felix Hoffman added the acyl radical ("acetyl") to ease the acidic effects that aspirin imposed on the lining of the stomach. Acetylsalicylic acid (ASA), known as Bayer's Aspirin, entered the marketplace. Since then, people have taken aspirin more than any other medication. Annual consumption in the United States has been estimated at thirty billion tablets.[6] Despite its long-term use, it was only in the 1970s that we came to understand aspirin's true mechanism of action. Aspirin blocks formation of prostaglandins, which are chemical messengers in the cascade that is the human body's reaction to inflammation or infection. Prostaglandins participate in stimulation of nerves, leakage of fluid from blood vessels into injured tissue, and elevated body temperature. Disrupting their formation therefore prevents pain, swelling, and fever. Prostaglandins also make platelets adhere, thereby assisting blood to clot. So, aspirin is effective at preventing the blood clots that might otherwise lead to heart attacks and strokes.

One might argue that no example is better than aspirin to illustrate humanity's debt to biodiversity. However, one challenger can give the willow tree a run for its money. We begin this story in the early seventeenth century in colonial South America. Agostino Salumbrino was an Italian Jesuit monk posted to the Spanish mission at San Pablo in the imperial town of Lima, Peru. An inquisitive and determined man, he was placed in charge of the mission's infirmary, but he soon realized this small clinic was no match for the disease burden of a burgeoning colonial city. The Jesuit philosophy placed a premium on spirituality in the natural world. In this vein, Father Salumbrino set out to create a *botica*, or pharmacy based on the abundance of natural remedies in the South American highlands. His apothecary would serve not just the needs of the local community, but the entire Viceroyalty of Peru. He tapped into a local culture rich with medical lore and commenced to plant a garden of plants with known medicinal properties, such as camphor, rue, nicotiana, saffron, and cana fistula, the last being wild cane used to remedy stomach disorders.

One can only imagine life in early seventeenth-century Lima. Disease was rampant, and as believed by Lima's inhabitants, afflicted persons without explanation. Among these diseases was malaria. Few diseases have brought more suffering to humanity. Hippocrates attributed malaria to the bad air, or *malus aria*, of fetid water. Scientists learned that the disease vector is not the swamp air but, rather, a swamp denizen—the *Anopheles* mosquito. This insect carries the deadly agent of malaria, a single-celled protozoan parasite of the genus *Plasmodium*. First, an infected female mosquito bites a human. While the mosquito is feeding, malaria protozoa enter the human bloodstream via the insect's saliva. The protozoa travel in the bloodstream, eventually reaching the victim's liver, where they take up residence and multiply. Infected liver cells soon burst, releasing millions of protozoa into the bloodstream. They invade red blood cells, in turn causing them to burst and release exponentially high numbers of the protozoa into the body. This recurrent cycle leads to a cascade of symptoms: fever, chills, headaches, muscle aches, jaundice (yellowing of the skin and eyes), vomiting, and diarrhea. Destruction of human red blood cells can cause rapid and severe anemia. Untreated, this painful and debilitating disease often leads to profound low blood pressure (contributing to "shock"), coma, and death.

Father Salumbrino had his work cut out for him caring for so many persons ill from malaria, but he was assisted by the wisdom of the indigenous people. Locals native to the Amazon region had long used bark of the cinchona tree to treat fevers and relax muscles. The bark contained quinine, the key ingredient for antimalarial drugs still used today. Jesuits in the field, particularly in the Loxa region northeast of Lima, began to collect the bark. Residents taught the Jesuits how to remove the bark in vertical strips so as not to kill the trees. Legend has it that the wife of the viceroy, the Countess of Chinchón, was one of the first Europeans to be cured of malaria by "Peruvian Bark."[7] The celebrity cure made the drug cinchona famous, and for more than a century, Europe's sole source of quinine was Father Salumbrino's dispensary in Lima.[8]

In 1820, French chemists isolated quinine from the bark of the cinchona tree, which led to synthesis of two antimalarial drugs, chloroquine and mefloquine.[9] By modifying these and other compounds, scientists have developed even better antimalarial agents, constantly striving to stay ahead of emerging drug-resistant strains of malaria.

The dramatic fashion in which quinine emerged to keep malaria at bay was eclipsed by a discovery in the jungles of Vietnam. After centuries of quinine-based therapy and many generations of mosquitoes selectively bred to resist quinine, the malaria parasites of Southeast Asia developed stubborn resistance to chloroquine, with devastating effects on the local human population. In the mid-1960s, this included the Viet Cong military. Desperate to shield his jungle-based troops from this scourge, Ho Chi Minh needed help. He turned to doctors in China's Red Army for help.[10] Chinese doctors began a top-secret program to develop new antimalarial treatments. Part of this effort instructed scientists to examine closely the value of Chinese herbs and folk medicines. Tu Youyou was a researcher at the Academy of Traditional Chinese Medicine assigned to the project. After screening more than two thousand traditional Chinese remedies, she came across a preparation described in a sixteen-hundred-year-old text titled "Emergency Prescriptions Kept Up One's Sleeve."[11] It described a treatment for "intermittent fevers," the defining symptom of malaria. She pursued the recipe and extracted artemisinin from the leaves of *Artemisia annua*. The results were miraculous. Artemisinin cleared malarial parasites from humans faster than any drug in history. China's harvesting artemisinin from the sweet wormwood plant revolutionized the treatment of malaria. For her work, Youyou received the Nobel Prize for Medicine in 2015.

One good thing can lead to another, so the situation might improve, because biochemists continue to explore the potential of artemisinin. The key curative mechanism of this drug is its interaction with iron, which is plentiful in malaria parasites. In this respect, protozoa are similar to cancer cells. Both the malaria parasites and cancer cells fuel their growth and division by using essential iron. Given that certain cancer cells, especially those of leukemia, show high iron concentration, artemisinin may be effective at killing them.[12] Initial studies of leukemia cells in tissue cultures have revealed tantalizing results using artemisinin as a therapy.[13]

The annals of ethnobotany provide us many lessons on the human health importance of biodiversity. Combining ancient wisdom about medicinal plants with contemporary scientific techniques will yield important discoveries only if plant species remain to investigate. If we remove the essential substrate, discovery becomes impossible. Modern science has only scratched the surface in a systemic approach to "bioprospecting," or performing deliberate research in biodiverse regions for new sources of chemical compounds with real or potential economic value. This is a slow process even if a plant is a likely prospect, because the successful pharmaceutical clinical trial as it currently is regulated usually takes more than a decade to complete.[14] Because we have rapid and widespread acceleration of biotic impoverishment, which is the destruction of biodiverse areas for the purpose of human use, conceivably we could run out of time for fruitful bioprospecting. This will be particularly difficult if more germs become resistant to antibiotics, in large measure because we feed antibiotics to livestock relentlessly. Consider the following case of methicillin-resistant *Staphylococcus aureus*.

GERMS WITHOUT BORDERS

Kayaking was her passion. Katie led groups of experienced paddlers down high-class rivers and through rapids that were the features of TV adventure documentaries. She loved every part of the process—mapping the route, scouting the river, renting the equipment, and, of course, listening to the shouts of triumph as her clients shot through churning whitewater between obstacles of boulders, jagged overhanging branches, and the occasional drop-off into a pool of swirling water. Occasional mishaps occurred, such as a turned ankle or bout of poison oak–induced dermatitis, but, knock on wood, nothing medically serious had ever ruined one of her trips.

Katie had heard other outfitters talk about a new kind of infection that followed minor scrapes with exposed rough edges of attachable seats in some of the open boats, the sea kayak–type vessels that could be used on tamer, slower water where less-experienced boaters could float without needing to perform technical kayaking maneuvers. A seemingly minor abrasion wouldn't heal, then would morph into something much more commonly associated with an infection seen in the arms of drug addicts who "skin popped" illicit substances into their forearms.

The scrape rapidly transformed into cellulitis, inflammation of the skin and tissues immediately underneath the skin. Healthy pink turned into angry red, with swelling and tenderness, creating a mound of angry germ-infected skin and fat headed for abscess formation. The culprit was methicillin-resistant *Staphylococcus aureus*, known as "MRSA." It wouldn't respond to treatment with available penicillins and cephalosporin antibiotics, and soon wouldn't respond to others. Unchecked by the miracles of modern medicine, pockets of pus could cause fever and highways of infection known as "blood poisoning" in the lymphatic system defended poorly by swollen lymph glands. If the circumstances and immune system of the victim favored more aggressive disease, a MRSA infection could spread rapidly in the body.

The new norm was never to share boat seats, lifejackets, towels, clothing, razors, or anything else that might be contaminated. Bleach was everywhere, and one didn't merely "wash"; he or she needed to "decontaminate." In a world where infections cannot be treated, prevention doesn't just eliminate a nuisance: it perhaps saves lives.

Medications still are available to treat MRSA infections, but over time, these may become ineffective. Since the beginning of recorded medicine, biodiversity has enabled scientists to discover new medications. If we allow global climate change to eliminate species, we most certainly will eliminate the possibility of attaining benefit from the plants and animals that are lost to us by our disregard for the environment and the fruits of biodiversity. Some scientists believe we may be in a period of mass animal and plant extinction.[15,16] If that is the case and untreatable infections emerge, will humans be far behind?

• *13* •

Ecosystem Services

$\mathcal{T}$he famous twentieth-century designer and inventor Buckminster Fuller once characterized our planet as "Spaceship Earth," reinforcing a perception that humans have everything they need to thrive in an otherwise hostile and desolate universe.[1] It is presumptuous to reach that conclusion, even more so to assume that even if it were true, nothing could change the situation. We evolved over millennia seemingly ideally suited to our environment. What if that environment changes?

If the environment is changing at a pace no longer measured in eons but in years, we need to take stock of the value of a stable and healthy ecosystem and understand the value imparted to human health by the concept of ecosystem services. These ecosystems account for the vast majority of our environment: grasslands, steppes, mountains, deserts, wetlands, estuaries, oceans, and so forth. When functioning properly, ecosystems provide material goods, such as agriculture, aquaculture, and woodlands; purify air and water; pollinate crops; decompose and recycle wastes; and regulate the climate. Disruption of these services can be catastrophic for humanity. For instance, disequilibrium in insect populations can cause widespread crop failures; the recent Colony Collapse Disorder of pollinators is a growing threat to food security.[2] The rise of carbon dioxide in the environment is hyper-energizing weather, acidifying oceans, and elevating global mean temperature.

These services exist on a scale so massive that it is not feasible to replace them. Likewise, we still have only a rudimentary understanding of what makes these services work, which species precisely are integral to these services, and in what numbers and proportions the species must exist for functionality.[3]

To attempt to understand these dynamic processes from a human perspective, environmental scholars have categorized ecosystem services into four major classifications: provisioning services, regulating services, cultural services, and supporting services. *Provisioning services* are the most intuitive ecosystem function, and they reflect humanity's historic perspective toward the environment (i.e., *what can be harvested for economic gain?*). Shelter, food, fuel, and commercial goods are byproducts of our ecosystems, so this perception contributed to the rise of human civilization through an understanding that manipulation of natural resources can produce crops and harness livestock. Provisioning services continue to underpin most of the economic productivity of the world's nations in the form of fish; livestock; dairy; wood; wool; fruit, vegetable, and cereal crops; and minerals. *Regulating services* keep our environment in balance and act as a cleaning service for human activities. Plants to some extent purify the air and, although slowly being overwhelmed, have acted as a considerable hedge against large increases in industrial particulate matter, as well as against nitric oxide and carbon dioxide.[4]

Wetlands and estuaries are excellent large-scale filters that purify water. Contaminated rainwater that percolates through soil is purified through chemical uptake by inorganic and organic compounds from plants and microbes, as well as being influenced by soil (clay and organic soil matrix). Nitrogen and phosphorous (both naturally occurring and from human-made fertilizer) are filtered out by wetlands and are in turn used by plants for growth. Soil microbes transform the water-soluble form of nitrogen into an inactive and harmless gaseous state. Estuaries (ecosystems where tidal waters meet inland rivers or streams) contain a host of aquatic life that somewhat clean the water. Mussels, clams, and oysters are excellent at consuming algae overgrowth (from excessive aquatic presence of fertilizer nutrients) and other suspended pollutants. One often cited example is the prodigious filtering capability of the eastern oyster (*Crassostrea virginica*). For years, these were so numerous in the Chesapeake Bay that they could filter its complete volume in just three days.[5] Today, through overharvesting, pollution, and habitat destruction, this same filtering mechanism takes approximately one hundred times longer.[6]

Another key regulating service of ecosystems is their ability to mitigate floods and control erosion. Floodplains are one of nature's safety valves. Floodplain ecosystems evolved along the great rivers to absorb water overflow from heavy rains and have created some of the world's most productive, nutrient-rich soils (Tigris and Euphrates, ancient Mesopotamia; Nile, Egypt; Missouri and Mississippi, United States). However, human activities often have disrupted this natural overflow, causing widespread destruction because floodwaters have no natural repository into which to spill. This is all the more concerning given the increasing incidence of climate-energized extreme weather events.

Three distinct human activities are attributed to a decline in aquatic regulating ecosystem services: widespread draining of floodplain wetlands, construction of permanent structures on wetlands, and building of system levees. Levees that are overwhelmed by flooding result in upstream flooding that surges into less-protected areas. This was the case during Hurricane Katrina, when the combination of levees and absence of adequate absorbing floodplain land led to the extreme floods witnessed in New Orleans.

Cultural services embody the emotional, spiritual, recreational, and psychological relationships that humans have with their environment. Across all cultures throughout history, nature has been an intimate part of the human experience, conferring a sense of place, imparting aesthetics, and stimulating all forms of expression that include the visual arts, music, language, and architecture. All cultures have recognized the rejuvenating effects of time spent in "nature." The opposite is also true. The psychosocial detriment of polluted and destroyed landscapes has well-documented negative effects on nearby populations.[7,8]

Last on the list is *supporting services*, which underpin the ability of all other ecosystem services to function. Examples of these services include nutrient cycling and pollination/seed dispersal. A handful of key elements sustain life on Earth, and ecosystems are the key recyclers that ensure availability to biological cycles. Carbon, nitrogen, phosphorous, oxygen, hydrogen, and dozens of other elements in inorganic and organic forms cycle through ecosystems and through bacteria, plants, animals, air, and water. A readily apparent life-sustaining cyclic process in ecosystems is the relationship between flowering plants and pollinators. Healthy ecosystems need nectar-producing plants to attract pollinators, which need to be in sufficient abundance to sustain the ecosystem. The recent demise of large populations of bees threatens sustainability of ecosystems and is very concerning to humans that rely upon their activities. In a similar manner, seed dispersion relies on animals to eat fruit, and roam. As natural habitats constrict and more species become subject to extinction, plant species also are at risk.

One way to measure supporting services is through the metric of *net primary production* (NPP), which is the amount of annual plant growth achieved through the process of photosynthesis, which is the ability of a plant to convert sunlight into energy. NPP is a measure of the potential energy of an ecosystem's ability to power and support all ecological processes, which thereby supports all other ecosystem services. It's also a surrogate metric to assess the impact of human activities upon a given ecosystem. Current studies estimate that humans consume or otherwise perturb roughly 40 percent of all planetary NPP, thus co-opting the potential energy of our ecosystems through activities such as construction, water diversion, and agriculture.[9,10] When considering ecosystem services as a whole, this has tremendous implications for net biodiversity. To put it another way, human land and water use have preserved species deemed valuable to humanity to the detriment of countless other ecosystem species. Much like the situation with climate change, we often don't understand the implications to our health until the damage is done, changes are entrenched, and it has become very difficult to modify or reverse what already has occurred.

CLIMATE CHANGE AND ECOSYSTEM SERVICES

Not surprising, climate change undermines global ecosystem stability. Increased mean temperature, extreme weather events, rising sea level, and elevated atmospheric and ocean carbon dioxide pose grave threats to the permanence of land, freshwater, and marine ecosystems.[11] Ecosystems are integral in buffering effects of extreme weather conditions on the movement of materials and flow of energy through an environment.[12] Climate change will increase vulnerability of ecosystems from extreme events and reduce their natural capacity to mitigate the impacts of such events.[13] Salt marshes, reefs, mangrove forests, and barrier islands provide the service of minimizing the impact of storm surges upon coastal ecosystems and infrastructure. Climate change threatens these natural features through sea-level rise and erosion from extreme precipitation. In the northeastern United States, even a small (1.6 feet) sea-level rise will dramatically increase the number of people (47 percent increase) and property loss (73 percent increase) affected by storm surges on Long Island compared with present-day storm surge impacts.[14] Heavy rains can diminish the adaptive capacity of ecosystems to handle pollutants. They reduce the absolute amount of time water can interact with reactive sites, through uptake or adsorption, and by washing away or injuring plants and microbes that remove pollutants from water.[15] When considering fertilizer use and heavy rains, ecosystems have been overwhelmed and become no longer able to regulate the downstream nitrogen balance. The effects of this phenomenon upon the Gulf of Mexico have been devastating. It is now the nation's largest hypoxic (low-oxygen) "dead" zone. Recent increases in rainfall in several regions of the United States have led to higher nitrogen amounts carried by rivers.[16,17] Climate data from the past fifty years show that the Mississippi Basin yields an additional thirty-two million acre-feet of water (equivalent to four Hudson Rivers) each year that is laden with materials washed from its farmlands.[18]

Droughts and warmer summer temperatures have increased the wildfire risk to historic levels, resulting in unprecedented social and economic challenges. In 2011, more than eight million acres burned in wildfires, causing fifteen deaths and more than $1.9 billion in property losses.[19]

Climate change is projected to reduce the ability of ecosystems to supply water to some parts of the United States. Compared to 10 percent of counties today, by 2050, 32 percent of counties will be at high or extreme risk of water shortages.[20] This will be most notable in arid parts of the country. In the southwestern United States, where the equivalent of 76 percent of all renewable freshwater is used by people for one reason or another, climate change–related drought effects (paucity of snowpack and spring precipitation) likely have diminished the quantity and altered the timing of spring stream flows, contributing to habitat loss and local extinctions of fish and other aquatic species.[21,22] In one projection, 47 percent of the trout habitat in the interior western United States would be lost by 2080.[23]

Some of the most obvious changes in ecosystems are occurring at the boundaries between biomes (i.e., regional ecosystems), as warming temperatures push historically warmer-habitat species higher in latitude and elevation. Examples include the boreal forest/tundra boundary in Alaska; subalpine forest/tundra boundary in the Sierra Nevada; broadleaf/conifer boundary in the Green Mountains of Vermont; shrub land/conifer forest boundary in Bandelier National Monument, New Mexico; and temperate mixed forest/conifer boundary in southern California.[24,25] Species will need to adapt or risk eradication from their historic habitats. It is projected that salmon, trout, whitefish, and char will be lost from lower-elevation streams in the United States, paralleling the trend in the ocean, where warm-water fish communities have displaced cold-water species.[26,27] Some species will adapt in a far more efficient manner to the extent that they will dominate an ecosystem, becoming "invasive" to that environment. For instance, the yellow starthistle plant (*Centaurea solstitialis*), a wildland pest that is predicted to thrive with increased atmospheric carbon dioxide, currently costs California ranchers and farmers $75 million per year in water losses.[28,29] Elsewhere, bark beetles have proliferated unchecked in extensive areas of the western United States and Canada, ravaging conifer forests in the worst outbreak in the past 125 years.[30] Warmer temperatures have been implicated as a major factor, allowing more beetles to survive winter, expanding breeding seasons, and increasing their historic range to higher elevations and latitudes.[31] The resultant "beetle kills" that destroy enormous acreages of forests leave dead trees that are quickly consumed in forest fires. Although some of this may be attributed to natural cycles, the warming may be premature and alter what might have been a less aggressive pattern of fires and habitat destruction.

Climate change threatens to extinguish species permanently. Iconic species, such as the polar bear and ringed seal, increasingly are under threat from warmer temperatures and degradation of ecosystems that sustain their health.[32] Such loss of biodiversity is a tragedy, because it is irrevocable and unnecessary.

LOSS OF BIODIVERSITY: THE CASE OF ARGENTINE HEMORRHAGIC FEVER

Ecosystems provide checks and balances. For instance, mosquitoes spread human diseases, but in a healthy ecosystem, many predators feed on mosquitoes, keeping epidemics from initiating or growing beyond containment. However, when a health-related ecosystem is stripped of its biodiversity, we can lose the natural defenses against disease vectors and run the risk of an unchecked disease situation becoming an epidemic.

Consider the Pampas of Argentina, one of the few "breadbaskets" of the world. This rich, fertile grassland encompasses more than three hundred thousand square miles. In the 1950s, the farmers of the Pampas made an effort to overemphasize farming corn for financial benefit, to the detriment of crop diversity. This monoculture of a single maize species reduced biodiversity of the ecosystem that had previously been lush with other crops. As the variability of the ecosystem decreased, so too did the animal inhabitants associated with each of the abandoned crops. One unintended consequence was that the massive corn output became a utopian habitat for the drylands vesper mouse (*Calomys musculinus*). These mice proliferated and became the dominant rodents in the region while other native species diminished.[33]

Unbeknownst to humans at this time, these mice were the natural animal reservoir for the Junin virus, which causes Argentine hemorrhagic fever (AHF) in humans. After human exposure to the virus through skin contact or inhalation, the virus incubates in the body for just under two weeks. The first symptoms then appear. These include fever, headaches, weakness, and loss of appetite. The syndrome progresses to a hemorrhagic (bleeding) state. Skin bruises are accompanied by gum bleeding and, sometimes, leaky blood vessels that may create fluid accumulation in the lungs or even circulatory collapse (shock). AHF carries a mortality rate above 30 percent.[34] As the mouse population grew, the virus came into much more frequent contact than previously with humans, causing increased exposure and subsequently major outbreaks of AHF.[35]

What is the lesson here for human health? It is that climate change, combined with human decisions based upon the outcomes of that change, has the potential to change patterns of plant and animal species, which in turn can promote situations of disease proliferation. It is not sufficient to take the most apparent observations and assume either the best or the worst. It is necessary to perform dynamic, in-depth expert analyses of situations that include coordinated input from environmental scientists, agronomists, botanists, zoologists, epidemiologists, and economists.

Part IV

SUMMATION

• 14 •

Climate Justice

$\mathcal{T}$he concept of *climate justice* enables a viewpoint that allows us to shape the conversation that should occur between policymakers and the public. This concept frames the impacts of climate change as an ethical issue seen through the lens of human rights. Anchoring this approach is the 1948 United Nations *Universal Declaration of Human Rights*, Article 25, which cites that "everyone has the right to a standard of living adequate for the health and well-being of self and family, including food, clothing, housing and medical care and necessary social services, and the right to security in the event of unemployment, sickness, disability, widowhood, old age or other lack of livelihood in circumstances beyond one's control."[1] The contention is that climate change directly undermines these rights, most demonstrably of the most vulnerable populations. By adopting this approach, proponents seek to link existing language and policy frameworks of global human-rights constructs to climate change and its effects, thereby using existing human-rights protections in laws and international agreements to facilitate climate change mitigation.

Mary Robinson, the former United Nations high commissioner for human rights and a thought leader in this field, explained:

> Climate justice highlights the impacts of climate change on the vulnerable, marginalised and poor, who are disproportionally affected by extreme weather events, such as floods, storms and droughts, and slow onset events, such as sea level rise and glacial melt. Climate change is already undermining many of their basic human rights—to food and water, to shelter and health. Climate justice points out that the undermining of these essential rights is an injustice . . . largely because those who are affected by the negative impacts are least responsible for the causes of the problem.[2]

The Human Rights Council, an intergovernmental body within the United Nations, passed a resolution in 2009 on human rights and climate change stating that climate change impacts the effective enjoyment of human rights, including the right to:

- life,
- adequate food,
- the highest attainable standard of health,
- adequate housing,
- self-determination, and
- human-rights obligations related to access to safe drinking water and sanitation.

The resolution recognized that the effects of climate change will be felt most acutely by segments of the population that already are in vulnerable situations owing to factors such as geography, poverty, gender, age, indigenous or minority status, and disability.[3]

This approach to the intersection of climate change and health is controversial, because it squarely holds governments and international institutions responsible for the effects of climate change and by logical extension, potentially liable for human-rights violations. It also implies that nations that have contributed and are now contributing to the largest sources of greenhouse gas emissions should be held accountable and therefore tasked with helping those who need to be helped. The concept of climate justice has arisen as a manifestation of the failure of governments to respond effectively to climate and climate-related health issues.

Although nations differ on the extent of a national health-care system, the right to the best possible health is an international norm. It is well known that the most socioeconomically vulnerable nations are the most likely to have deficient health-care systems and therefore frequently struggle to provide adequate health care. These same health systems will be challenged with greater demands arising from an increased disease burden from waterborne and vector-borne pathogens and other climate-related illnesses derived from extreme heat, poor nutrition, and unsafe air quality. As more people succumb to malaria and Zika virus outbreaks, understaffed clinics, already laboring under the load, will struggle to keep up with additional patients. In this regard, the health rights of the people who already are suffering from poor baseline health, limited education, and inaccessible health services are most at risk for further deterioration.[4]

THE MOST VULNERABLE

From a human-rights perspective, we can assess vulnerability in many ways. Geography plays a role. For example, the 3.2 billion people who live in coastal regions worldwide will be at risk from floods and storm surges created by sea-level rise.[5] Preparedness will mitigate the hazard to some extent. The citizens of a well-resourced city such as New York City have a much different risk profile than persons living in Dhaka, Bangladesh. When it comes to human-rights vulnerability, nothing correlates as closely as does poverty. The World Bank has set the poverty definition to be persons earning less than the equivalent of $1.25 per day. Some 1.2 billion people worldwide meet this definition.[6] The United Nations Development Multidimensional Poverty Index, which measures the intensity of poverty based on deprivations in basic services of education, health, standard of living, and core human functioning, sets higher the number of impoverished people, at 1.7 billion people worldwide. More than 70 percent of persons in poverty reside in lower- and upper-middle-income countries, which are the same nations that have fewer resources and frequently poor governance.[7] Climate-related events will expose the vulnerabilities of the poor through disruptions in economic productivity, threats to agricultural yields with higher likelihoods of food price spikes, damaged homes, and loss of a sense of security and well-being. Poverty traps, which are the cyclical deprivations that often pass from one family generation to the next, can become embedded by extreme weather events that render families unable to rebuild lost assets.

Demography is another major determinant of human-rights vulnerability. The global elderly demographic is growing faster than any other age group and is estimated to make up 16 percent (double the current amount) of the world's population by 2050.[8] Elders have less physiologic reserve and are more likely to have disease comorbidities (heart disease, diabetes, lung disease), which confer greater susceptibility to health stressors. Older people often depend on family caretakers, which means that their health situation is in the hands of others. In 2016, simply being female is still considered a health risk factor. Marginalization of women is prevalent in much of the world, with limited or even complete denial of access to education, economic opportunity, and mobility. Children are vulnerable for obvious reasons related to age. Climate change is not under any of their control, so if it erodes health, they are defenseless.

Perhaps the greatest health rights vulnerability from climate change is the injustice perpetrated within a span of time that doesn't reach its full effect until considerably later. This notion calls attention to the fact that our actions or omissions today will affect future generations more profoundly than they will our generation. It is very possible that our failure to halt human-caused climate change will not show itself until the next generation, or future generations. University of California Berkeley professor Kirk Smith points out that "damage to others, including from climate change, can be propagated across time by our own present actions, even some actions that seem rational or apparently harmless in the short term."[9] The moral implication of this "intergenerational equity" holds our contemporaries and us liable for human-rights violations against future generations. Perhaps, like never before, they may suffer for our failures.

THE FULCRUM

One of the most compelling arguments for a rights-based approach is that it can rely on existing covenants and agreements, and reinforces the imperative for long-term solutions. By invoking legality and internationally accepted ethical norms, advocates of climate justice seek to mobilize the international human-rights regime in which stakeholders have a preexisting duty to uphold these rights, irrespective of sovereign considerations.

On paper, the duties of international states and partners are clear. Subject to adequate resourcing, states are expected to offer financial and technical assistance to less well-resourced states in order for them to realize human rights effectively.[10] In the context of climate change, it codifies a state's responsibility to control the impact of its climate-related actions and policies not only upon its own citizens but also upon other people around the world. This can be the basis upon which developing countries ask for international assistance to support climate change mitigation and adaptation or response plans. If a state's decisions or actions are deemed to violate human rights by exacerbating climate change impacts upon human health, then an obligation exists to fix or enact policies to change this pattern. In the spectrum of human rights, every nation and, indeed, every individual is held accountable for their actions.

Of course, this is nearly impossible to enforce. However, this strategy has had some philosophical success, especially with low-lying island nations whose citizens are most at risk from sea-level rise. In September 2007, the Maldives asked the United Nations Human Rights Council to hold a debate on the link between climate change and the full enjoyment of human rights. The meeting took place at the Small Island States Conference later that year, and the outcome was the Malé Declaration on the Human Dimension of Global Climate Change, which for the first time demonstrated an international agreement that explicitly stated, "climate change has clear and immediate implications for the full enjoyment of human rights."[11]

A human-rights-based approach to climate change is a means to an end.[12] In realizing basic human rights through the principles of equality and non-discrimination, accountability can be infused into the policy discourse, along with a moral imperative to act for the preservation of human health worldwide. We should be planning for the future not only as if our lives depended upon it, but knowing full well that the lives of forthcoming generations depend upon it. Medical justice demands this approach.

• *15* •

Rx for the Future

After two weeks in the intensive care unit, there was little left of Sid. His emotional resources and will to live were gone, so his family requested that the breathing tube be removed. Without machines to support him, he would receive "comfort care" only. Sid and the family at his bedside knew that his luck had run out.

"How much time does he have?" his daughter asked.

The palliative care specialist was gentle and honest. He simply shook his head side to side. Sid's prognosis was grim, and the end was near. The silence was more powerful than words.

His daughter knew that Sid had spent many years abusing his health. She looked away and with a mournful tone whispered, "Wish we could go back forty years and show him how this would all end up. He probably knew what was coming. We just couldn't change him. I love you, Dad."

We are doctors who care for our patients, and the planet that is their home. We wrote this book because we believe that humans are making changes to our environment that soon may become irreversible. We fully understand that the science of climate change is evolving as experts make more observations, perform more research, derive more data, and do their best to make reasonable interpretations. The controversy is apparent, but the logic of preserving what we can and trying to avoid depletion of resources and immutable changes to our planet that will affect human health adversely are truths to us. We took a "doctor's approach" because we believe we are facing a sick patient, and we want emphatically to bring human health to the top of the discussion. We presume it is highly likely that humans are altering the environment in a way that causes global warming, widespread pollution, destruction of habitats, and everything that comes with these. From that assumption come logical human health implications, and you are entitled to know about them.

We care about you. If your chest was hurting, you couldn't breathe, and your pulse was undetectable, we wouldn't sit still and ponder the situation. We'd do everything possible to make an accurate diagnosis and try to save your life. We'd act fast, because we know that moments count. Should we be any less concerned about our planet? The current situation in Glacier National Park is instructive. In our children's lifetime, it may need to be renamed Glacierless National Park. Does that matter? We believe it does.

We rely upon the science of others to understand predictions of global climate change, but we do not need others to explain the health effects that logically would be related, because these are conditions we know. We treat them. The composite patients suffering from post-flood diarrhea, wildfire-induced shortness of breath, and heatstroke-induced organ failure are the same cases seen every day in emergency departments and medical wards around the globe. Climate change will add to the burden of disease, and sooner or later it will affect people *you* know, and hundreds of millions you never have met.

Part of the problem to date with science communications on climate change has been failure to identify an immediate health threat. Without that stimulus, the imperative to change the way we live and consume our resources is not sufficient. Lacking a perfectly accurate crystal ball is an impediment for some to become engaged in this issue. We want everyone to form an opinion, and to act on their knowledge and conscience.

If you disagree with our premise, then learn and respond. Throwing up your hands in frustration over the futility of our future will not solve anything. Consider recent events. Fifty years ago, 42 percent of Americans smoked cigarettes.[1] Although tobacco use was widely suspected to cause lung disease, the national per capita cigarette consumption had been on the rise since the nineteenth century.[2] Medical scientists pointed out the hazards. However, the industry sponsored efforts to suppress the science about detrimental health effects. Such was the state of affairs that some physicians themselves became spokespersons about the benefits of cigarette smoking.[3]

Today, only just over 17 percent of adults in the United States call themselves smokers.[4] How did this change? For one thing, the financial cost of smoking became prohibitive (economics). Tobacco smoking was banned from public places (regulation). A relentless public health media campaign (medical science) made it impossible to ignore the fact that smoking causes major health risks (public opinion). Although this may not be precisely how to affect changing attitudes and behaviors regarding climate change, it clarifies two things. When effects from a behavioral pattern can be determined and linked to a negative economic or health risk profile, public behavior can change. Furthermore, seemingly insurmountable situations that require support of the masses can be overcome.

We take inspiration from an analogous and present existential threat. During the 1980s, frightening escalation of the Cold War invoked rhetoric that nuclear wars could be "winnable." A group of visionary doctors founded the International Physicians for the Prevention of Nuclear War (IPPNW) and rendered a clinical assessment that was a sober rebuke to Cold War strategists. *Nuclear war would be the final epidemic; there would be no cure and no meaningful medical response.*[5] By the fall of the Berlin Wall, the IPPNW was two hundred thousand members strong and had played an instrumental role in promoting international agreements to ban nuclear test explosions and close nuclear weapons testing sites and production facilities. The source issues of nuclear proliferation were complex and involved conflicting science, national security concerns, and economic policy. Still, key factors led to the IPPNW's widespread appeal. Members spoke with unambiguous clarity in the face of dissonance from multiple sources, including governments, lobbyists, and political entities. Their position was unconditional and clear—the health risks of nuclear war were massive and irrevocable.[6] Sadly, because of recurrent nuclear weapons activities, soon we may be having these discussions once again. But at least we can agree that the danger exists and proceed on that basis.

WHAT ABOUT "GEOENGINEERING"?

We are of the opinion that causation exists for any demonstrable harm. From that vantage point, prevention is the key to the cure. But we also treat patients after they have become ill, and we do our best to not give up on them. Might the same approach work for planet Earth? Can we alter it and then accommodate its afflictions or nurse it back to health? We are not experts on these matters but note that a growing movement arising in certain scientific circles espouses this approach. The thesis is, if we can't curtail human activities to the point where planetary environmental changes can be stabilized spontaneously or reversed, then we should invest in technologies actively to stabilize or reverse them. This concept of *geoengineering* has gained traction as a way to minimize the seemingly inevitable trend toward a permanently altered world.

Two basic methods are integral to geoengineering. The first is carbon dioxide removal. Essentially, this would manipulate the global carbon cycle to extract carbon dioxide from the air and sequester it someplace else. Various proposals include storage in the soil, vegetation, oceans, and underground.[7] The second method is solar radiation management, which aims to counteract global warming by reflecting back a greater proportion of incoming solar radiation into space.[8]

This treatment approach is familiar to us as physicians. We often confront disease in its later stages, once illness has entrenched itself irrevocably. Although medical science has made great progress treating advanced heart disease, end-stage kidney failure, and invasive cancers, no doctor ever would suggest that a patient keep smoking because a new inhaler treats chronic lung disease. We know well that one key to optimal health is to avoid getting sick in the first place. We reject the notion that it's sensible to bioengineer our planet rather than attempt to eliminate the behaviors that ruin its condition, particularly when the predictive models inform us that we still have time to stabilize our climate and avoid the worst of these outcomes.

WHERE DO WE GO FROM HERE?

The first step is to clarify environmentally and medically what is happening. We need to agree on the science and implications. These are matters to be decided objectively, certainly not by persons who stand to profit economically or politically from the outcome of proposed actions.

The second step is to consider climate change in much larger part as a health issue so that we can understand fully what's at stake for long-term consequences to the human race. The prestigious medical journal *Lancet* framed climate change as the "greatest global health opportunity of the 21st century."[9] The multidisciplinary, international Lancet Commission put forth ten policy responses to ensure the highest attainable standards of health for populations worldwide. These include public health initiatives; investment in renewable energies, especially in low- to middle-income countries; advancing of carbon-pricing initiatives; strengthening global governance on climate accords; and meticulously monitoring the health metrics of global health and climate change.

The third step is to vote for motivated public officials. Responsible legislators should be held accountable for reading the science and expressing their opinions. Government doesn't move unless it's pushed, so let's work within the system to push it in the right direction. Many of us feel disillusioned by the political process, but that is what we have to work with. The most direct way for us to influence policymakers to address the issues that matter is to know our stuff, communicate with elected officials, and back our words with our votes.

The fourth step is to purchase wisely and promote environmentally preservative practices. Learn what you can about electric and hybrid cars; wind and solar electricity generation, and battery technology; environmental practices in your schools and businesses; and many other opportunities. You can steer toward a more environmentally responsible world by making choices in what you buy and how you consume.

The fifth step is to let yourself be heard. Use your voice, pen, and ballot. If you believe in change, then pursue it. Always remember that you are doing this for the children. We should give them a fighting chance to have a planet that will support them—and future generations.

The sixth step is to feel a sense of responsibility for your fellow humans. It is so easy to be provincial and let others fend for themselves. We have both been in situations where others needed our help in dire, difficult circumstances. We did our best to be of immediate assistance, but more important, to put something in place so that when we departed, we left behind something useful. That is the situation we have now. We need to make things better soon and create strategies that will enable the improvements to persist and grow.

THE FINAL DIAGNOSIS

If Earth is warming because of global climate change, then human health will suffer. If we are not destined to drive ourselves to extinction, it's time to come together as a unified human race and face this issue. We do not appear to be on a good path. The physical planet will be here, but we may not. If we need fertile fields to grow crops and livestock, oceans to supply precious chemicals and nutrients, clean air to breathe, and weather patterns that don't threaten large populations continuously, we need to make prompt decisions. Human history is replete with creativity, resilience, determination, and optimism. Let's get going.

Recall the paintings of Chauvet-Pont-d'Arc Cave in southern France, conceived more than thirty thousand years ago. Among the breathtaking drawings of horses, cattle, mammoths, lions, and bears, we see a single handprint. Our ancestors spoke to us: "We are here . . . and we are going to do great things." How will history remember us? Will we be the cursed generation of the era of the "Great Procrastination," or will we educate each other, change our behaviors, and protect our progeny with a likeminded purpose?[10] We hope it's the latter. If we succeed, perhaps our descendants will view us millennia hence with gratitude and respect because we embodied the best attributes of the human spirit and left them a legacy of creativity, peace, prosperity, and the best possible place to live.

Afterword

$\mathcal{N}$ow begins the most important work of our lives. We conceived the term *Enviromedics* as a unique and hopefully catchy moniker to describe the effects, consequences, and study of environmental change upon human health. But we also envisioned it as a word around which to rally calls to action for the profound and creative applied science and collective determination that will be necessary to undertake the Herculean effort to raise awareness and preserve human health in the midst of a changing environment.

Our inventory of adverse health impacts portends a harsher world. But let us be optimistic that bright minds and sincere young people will not sit still for such a predetermined future. Along with global conflict, this is an issue for the ages. It is multi-generational, and so as we have always done for our children, we must create and support brilliant innovation and leadership to tackle these problems. Economists will learn to lower costs and increase efficiencies for renewable energy sources; social scientists and venture capitalists will promote social entrepreneurism; and politicians will harness their constituents to recognize that we must have an eye to the future in every endeavor that might impact climate change and human health. We cannot just sustain. We must prosper.

This is our mission, but hardly ours alone. Future generations will largely deal with the health burdens we describe in this book. Many have already accepted the challenge, and our universities support the disciplines necessary for this task. Their students are not yet the decision-makers, but they will be, so now is the time to begin to give them a meaningful voice in the issues that will mold their future. As has been the case since the dawn of civilization, the hope lies in our youth.

We've invited accomplished members of the next generation to offer perspectives and reflect on *Enviromedics*. They represent many thousands of dedicated young researchers, inventors, entrepreneurs, and other thought leaders who have accepted the challenges posed by our current situation.

Jaden Smith has made a significant commitment toward entrepreneurism in renewable energies and resource efficiencies. He is a co-founder of JUST water, a bottled water company designed to ethically source water, reduce fossil fuel use and greenhouse gasses emissions in packaging, and direct money back to community infrastructure repairs. He is a co-founder of a water reuse and emergency response company, and co-founder of JUST Impact, a nonprofit organization focused on developing solutions to energy, food, and water challenges. He will launch an innovation laboratory at the Massachusetts Institute of Technology (MIT) in 2018 with the intent to help identify and support groundbreaking engineering in the fields of Energy and Water. *From Jaden:*

"Raising awareness of my generation to the environmental threats caused by climate change and supporting those working to mitigate them is a personal passion of mine. I'm getting up to speed as fast as I can, but when I read *Enviromedics*, I realize that there is an entire dimension to this crisis that we've hardly emphasized. My eyes opened wide when I learned about the very serious effects climate change can have on people's health around the globe—not in fifty years, but right now.

Climate change has no borders. It affects everyone—old and young, rich and poor. We see its omnipresent and unavoidable consequences through sea-level rise, droughts, and extreme weather events. Now I also understand that climate change means that populations will have accelerating health issues like asthma, malaria, heatstroke, and cardiovascular disease. There it is, right in front of us. I ask the same question that Drs. Lemery and Auerbach ask in this book: Why aren't we paying more attention to the threats to our health and to the health of those we love? What are we waiting for?

I'm not a scientist or a doctor. I am, however, a passionate young person who will soon be an adult. You don't need 'MD' next to your name to care about your personal health and the well-being of others, or be a climate expert to care about the environment, to learn, and to take responsible action. I am a devoted supporter of environmental solutions not just for my family and friends around me, but for my children and people whom I will never meet in my lifetime—the generations that will follow us.

I am writing this because the problems stated in this book will have to be solved by our generation. Yes—*OUR generation*, perhaps the last one that can have a serious impact before we do damage to our planet that cannot be reversed. The details about climate change and its effect on our health have not been on the forefront of the minds of millennials. But they should be. This issue could accelerate and become more catastrophic and less reversible each year we don't become involved.

It will be our generation and beyond that is forced to redesign our lives in Miami, Venice, Tokyo, Boston, New York, Bangkok, Manila, Shanghai, and countless other cities worldwide. Not only might we have to redesign these places at a massive economic cost, but we will have to do this while facing unprecedented physical and mental stressors associated with human displacement and migration, with many of us assuming the burden of what unfortunately will become a common label—climate refugees.

This is my call to action for everyone of my generation. Now is the time to discuss the effects that climate change has on human health. Now is our time to get a head start on what might otherwise become an unavoidable future. Learn as much as you can. Work hard in school and appreciate all sides of every issue. Go into science and engineering to help steer your mind and energies into clean technology, conservation, and environmental studies and law. Join and support policy advocates like the Natural Resources Defense Council. Contact your local political representatives and speak your mind about their role in supporting climate mitigation laws to protect our air and water, and our future. Go into medicine with an understanding that health-care professionals will be in the thick of the fray. Be bold. Start conversations. Speak up. Vote. Build a sustainable life for yourself and the people around you.

Thank you to Drs. Lemery and Auerbach for letting me help them bring *Enviromedics* to my generation. We need to follow the best role models to become role models ourselves. Our parents are doing this, and so must we. We're up to it, at the forefront of being able to uniquely affect human history by saving our planet, so let's get busy and do it."

Charles Orgbon III is a recent college graduate. He started Greening Forward when he was twelve years old. Devoted to providing the tools and resources needed by youths to create a sustainable world, the organization is one of the largest entirely youth-driven nonprofit organizations. Its "Guiding Beliefs" includes, "We don't wait until youth are 'grown-up' to teach them how to read and write; we also should not wait until they are 'grown-up' to equip them with the tools that they need to make a positive environmental difference in their communities. When youth engagement and youth voice are substantively added to strategy, transformative things happen within our communities and schools." *From Charles:*

"As a fifth grader, I went to a school that required community service. Like my peers, I was looking for the easiest thing I could do to get an 'A,' so I began looking at the issues in my community and noticed my school's littering problem. I was disturbed and wondered why anyone would do something so blatantly wrong for our planet, so I told myself and everyone who would listen, 'I'm going to free the world of litter.' To learn more about the issue, I went to the internet. The almighty Google had the answer. I soon realized that there was far more to environmental damage than simply littering. Our oceans are under siege, and wild habitats are being destroyed. I felt that it is our responsibility to do something about this. I started a website to share my thoughts on environmental issues. What I first saw as an opportunity to be creative and have fun became the beginning of a global movement that sought to change the way young people engage in pressing challenges, such as climate change.

I found other young people, and we launched Greening Forward, creating a network of young people who were eager to do something about environmental issues. We worked with funders such as singer-songwriter Jack Johnson, LUSH Cosmetics, Ben & Jerry's, and Levi Strauss to be able to direct more than $70,000 in funding to youth-inspired environmental projects. These youth planted hundreds of trees, diverted tons of waste from landfills, organized international youth summits, and completed dozens of other projects that conserved natural resources and advocated for sensible environmental policies.

To me, this fervor from young people is not surprising because my generation is likely the first to feel the growing effects of climate change, pollution, and ecological degradation within their communities.

I am acutely aware of the impacts not only upon the physical environment but also upon the people who bear its burdens. I was in my final year at the University of Georgia when severe forest fires hit the northern part of our state and spewed clouds of smoke onto campus that lingered for days. I had an acute allergic response to the smoke. At the time, I was preparing to leave the country for Santiago, Chile, where I was planning to help bring Greening Forward's environmental leadership materials to the schools there. I went to the doctor to ensure that my health was sufficient for the trip. The doctor assured me, "Wait until you're in Santiago. Your condition will improve there." Upon disembarking from the plane in Santiago, I discovered that Chile was in the midst of breaking records for forest fire incidents. Similar to what I had just experienced in Georgia, smoke shrouded the air, and I recalled how scientists tell us that because of climate change, some parts of the world will be drier, and, as a result, forest fires will occur more frequently. I was also reminded that climate change is global, and inescapable.

More so now than ever before, we need a population of young people who not only learn traditional education topics, but who also are taught how to solve global, complex problems such as climate change mitigation and adaptation to possible environmental conditions. We are up to the challenge. I am honored to be among the generation that is mobilizing to be part of the solution."

Brianna Fruean is a teenage environmental advocate for Samoa. As a founding member of 350.Samoa, she participates in projects to help combat climate change and promote sustainable development. She has advocated for her country and region at many international conferences, and at age sixteen became the youngest ever recipient of a Pacific Region Commonwealth Youth Award. She is a youth ambassador for the Secretariat of the Pacific Regional Environment Programme. *From Brianna*:

"Growing up on a small island in the middle of the South Pacific Ocean, 'climate change' is a term I know too well. As a five-year-old, all I could understand was that it was this big evil thing that would make my island hotter. When I turned ten, I understood that it was the cause of extreme weather patterns that would uproot trees and flood my grandma's house. When I turned fifteen, I understood why my neighbors in Tuvalu were purchasing land in Fiji, because they could not hold back the ocean from engulfing their land. My parents taught me many things about climate change, like 'climate adaptation' and 'ocean acidification,' that they were trying to learn themselves. My grandfather often speaks about fishing with my father when he was a boy—the cool ocean breeze, the vibrant colors of the reef, and the abundance of all kinds of fish. That experience is becoming rare.

Since the birth of our islands, the ocean and village plantations have been the main source of food for our people. In recent years, we have seen a massive drop in fish stock. Escaping the heat or searching for healthy unbleached coral has fish migrating farther from the reefs and into deeper water, making them harder to find. Our farmers struggle with changes in weather patterns that affect their harvests. They don't know the right time to plant which crop because there are floods during dry season and drought during rainy season. Natural food security is declining at a frightening rate. As a result of climate impact, more families buy canned fish, packet noodles, and other processed foods that do not provide the essential nutrients of a traditional diet of fish and taro. So, if our natural environment is unhealthy, it directly affects our own health. If it's bad now, we can only imagine what will happen to the health of future generations.

Samoa is also an example of extreme weather patterns intensified by climate change. In 2012 I experienced the wrath of Cyclone Evan. Following the cyclone, diseases such as dengue fever and conjunctivitis started spreading like crazy. I caught both and was sick while I had to sweep the mud from the flooding out of my house.

My generation may have inherited a damaged planet, but time and time again I have witnessed young people fight to save it. I've seen youth lobby policymakers, replant coral, clean shorelines, invent green technology, sue their governments, and spread the message that the chain of damage will stop with us. There's no easy bandage for climate change, but one thing is for sure: the damage done by the generations before us can only be fixed through the cooperation of several generations today. All hands must be on deck.

In Tokelau, a tiny island in Polynesia, the best navigating canoes were most efficient because they had a common setup. The person at the back of the canoe is the Ulu Hina (Grey Hair) who navigates the vessel; this is a person who has knowledge and respect in the community. While the canoe is guided by the elder, it is powered by the young people in the front, who paddle with passion. This is my favorite example of an intergenerational partnership. Although the world is ruled by older (mostly) politicians, what group makes up one of the biggest percentages of people on Earth? Who will work, live, create, and heal? Our young people! Tomorrow's future leaders are at least as important as today's. Intergenerational collaboration in climate invention and action is of utmost importance. The technology we invent today, negotiations we make today, and course for the environment we chart today will define the future for generations to come. We are committed, and we should be heard."

Xiuhtzcatl Martinez is a teenage environmental activist and youth director for Earth Guardians while also maintaining a related career as a hip-hop artist speaking to the state of the planet. He received the 2013 United States Community Service Award from President Obama and served on the President's youth council. Among many other awards he has received for being a spokesperson for environmental issues to young people are the Peace First Prize and Children's Climate Prize. *From Xiutzcatl*:

There is a misconception that people think of climate change as only affecting the health of our planet; *Enviromedics* changes that conversation by focusing on the impact upon human health and the imperative to change our behaviors and lifestyles.

The effects of climate change will bequeath a disproportionate impact upon indigenous communities and communities of color—places that lack access to resources of resiliency and available medical care. A few years ago, massive flooding from an extreme precipitation event hit my hometown of Boulder, Colorado. We were fortunate to have had resources to assist people who lost their homes and to ensure continued access for all to clean water. Imagine that same storm's impact had it struck rural Mexico. It would have been devastating. The health effects from the climate crisis brought on by continued fossil fuel extraction will transform entire regions, and we are not prepared for these changes.

I'm seventeen years old. I can't yet vote. The youth of the world are an underrepresented demographic. Of all people, we should be the major consideration for avoiding climate catastrophe. So, there is not intergenerational justice regarding climate change. The health burdens of the future are being imposed by the actions and policies of today. But while many spend enormous energy to decry this as an indictment upon our parents and grandparents, I see intergenerational *action* as our greatest chance for success. Our fullest potential is realized when we acknowledge strength in diversity. This lesson has been proven over and over again, through actions like the protests at Standing Rock, the student divestment movement, and the People's Climate March. Unified voices create a massive ability to inspire change. After all of our current politicians and policymakers are long gone, it is our generation that will leave its mark on this Earth and define a legacy that will protect our children and our children's children.

Caroline Spears is a master's student in civil and environmental engineering at Stanford University studying energy access and climate change. She is co-director of Students for a Sustainable Stanford, and has worked with the Environmental Defense Fund and as a legal intern for the Environment and Natural Resources Division of the U.S. Department of Justice. Caroline is particularly interested in how new energy technologies are deployed, and how they affect the world around them. *From Caroline*:

"My phone alarm rings, I take a breath of air, I eat a bowl of cereal. To do this, somewhere, thousands of people are working in a lithium mine that supplies the metals for each phone battery. Air particles—a mix of oxygen, nitrogen, and particulate matter specific to my location—enter my lungs. Milk from a dairy farm 500 miles away, mixed with a wheat crop in the Midwest, has been placed in my bowl in a dorm in California.

We take a lot for granted: that drinkable water will come from the faucet, or that affordable food will be available at the grocery store. In the past, I've also assumed that our elected leaders will care about facts, that our government will attempt to improve people's lives, and that the infrastructure investments we've made will not go to waste because of climbing temperatures or rising seas.

Coming of age in today's society has required some heart-wrenching realizations, and I no longer take for granted many things. I have lost faith in the idea that truth will trump political ideologies, and in the ability of some people to stay true to their own values in the face of uncertainty and stress. I no longer believe that the arc of the moral universe, however long, will automatically bend toward justice.

With these realizations I can either give up or be inspired. I choose the latter. The arc of the moral universe will bend toward justice because we will be there bending it. The will to distinguish facts from fiction will return to our political institutions only by the insistence of a people reconnected with democracy. I will stay engaged, and others will with me, because we refuse to accept that our society is irreparably separated from scientific truths that, as Drs. Lemery and Auerbach have shown us, have enormous consequences for our health and welfare. As a young person, I am worried. But I am also motivated by 'We the people'—a phrase that, I hope, includes you."

We know that these are not isolated voices. They are heartfelt and will hopefully help harness the passion for learning, discovery, discussion, and action across generations that have a duty to respond. We understand that our current health priorities have been centered on how we manage tobacco, prevent communicable diseases, diminish obesity, and mitigate the hazards of all manner of human conflict. We should now definitively add health matters to the discussion on global climate change, and elevate consideration of the human condition to where it belongs in that discussion—at the top of the list.

Notes

FOREWORD

1. Available online at http://classics.mit.edu//Hippocrates/airwatpl.html
2. A. Spielman and M. D'Antonio, *Mosquito: The Story of Man's Deadliest Foe*, 2001.

PART I

1. N. Watts, W. N. Adger, and P. Agnolucci et al. "Health and Climate Change: Policy Responses to Protect Public Health." *Lancet* 386, no. 10006 (2015): 1861–1914.
2. Centers for Disease Control and Prevention (CDC) Climate & Health Program. https://www.cdc.gov/climateandhealth/. Accessed March 16, 2017.

CHAPTER 2

1. D. Griggs. "Climate Policy: Streamline IPCC Reports." *Nature* 508, no. 7495 (2014): 171–73.
2. J. Goodell. "Global Warming Is Very Real," *Rolling Stone*, September 12, 2013. Available at: http://www.rollingstone.com/politics/news/global-warming-is-very-real-20130912. Accessed May 4, 2017.
3. Intergovernmental Panel on Climate Change. Fifth Assessment Report (AR5), https://www.ipcc.ch/report/ar5/. Accessed October 1, 2016.
4. National Ice Core Laboratory, http://www.icecores.org/about/index.shtm. Accessed October 4, 2016.
5. National Oceanic and Atmospheric Administration (NOAA), http://research.noaa.gov/News/NewsArchive/LatestNews/TabId/684/ArtMID/1768/ArticleID/11153/Greenhouse-gas-benchmark-reached-.aspx. Accessed October 12, 2016.

6. National Aeronautics and Space Administration (NASA), http://www.nasa.gov/feature/goddard/2016/climate-trends-continue-to-break-records. Accessed October 12, 2016.

7. United States Geological Survey, https://www2.usgs.gov/faq/node/3469. Accessed October 12, 2016.

8. J. A. Church, P. U. Clark, A. Cazenave et al. Intergovernmental Panel on Climate Change. Fifth Assessment Report (AR5); chap. 13, Sea Level Change, https://www.ipcc.ch/pdf/assessment-report/ar5/wg1/WG1AR5_Chapter13_FINAL.pdf. Accessed October 1, 2016.

9. R. W. Nichols, and H. H. Schmitt. "The Phony War against CO2." *Wall Street Journal*, November 1, 2016, A11.

10. O. Hoegh-Guldberg, P. J. Mumby, A. J. Hooten, R. S. Steneck, P. Greenfield, E. Gomez, . . . and N. Knowlton. "Coral Reefs under Rapid Climate Change and Ocean Acidification." *Science* 318, no. 5857 (2007): 1737–42.

11. National Centers for Environmental Information, https://www.ncdc.noaa.gov/sotc/national/201503. Accessed November 28, 2016.

12. NOAA, https://www.climate.gov/news-features/understanding-climate/2013-state-climate-record-breaking-super-typhoon-haiyan. Accessed October 12, 2016.

13. National Hurricane Center, http://www.nhc.noaa.gov/data/tcr/EP202015_Patricia.pdf. Accessed October 12, 2016.

14. Intergovernmental Panel on Climate Change. Fifth Assessment Report (AR5).

15. Frederick Burkle, "Foundations for Global Health Responders." Massive Open Online Course (Coursera, 2015), https://www.coursera.org/learn/ghresponder.

16. Intergovernmental Panel on Climate Change. Fifth Assessment Report (AR5).

17. Ibid.

18. M. Beniston. "The 2003 Heat Wave in Europe: A Shape of Things to Come? An Analysis Based on Swiss Climatological Data and Model Simulations." *Geophysical Research Letters* 31, no. 2 (2004).

19. A. Fouillet, G. Rey, V. Wagner, K. Laaidi, P. Empereur-Bissonnet, A. Le Tertre, . . . and E. Jougla. "Has the Impact of Heat Waves on Mortality Changed in France since the European Heat Wave of Summer 2003? A Study of the 2006 Heat Wave." *International Journal of Epidemiology* 37, no. 2 (2008): 309–17.

20. A. Zanobetti, M. S. O'Neill, C. J. Gronlund, and J. D. Schwartz. "Summer Temperature Variability and Long-Term Survival among Elderly People with Chronic Disease." *Proceedings of the National Academy of Sciences* 109, no. 17 (2012): 6608–13. doi:10.1073/pnas.1113070109.

21. World Health Organization (WHO), http://www.who.int/gho/urban_health/situation_trends/urban_population_growth_text/en. Accessed October 12, 2016.

22. United Nations World Urbanization Prospects Report, http://www.un.org/en/development/desa/news/population/world-urbanization-prospects-2014.html. Accessed October 12, 2016.

23. United Nations World Urbanization Prospects, https://esa.un.org/unpd/wup/Publications/Files/WUP2014-Highlights.pdf. Accessed November 28, 2016.

24. S. L. Harlan, A. J. Brazel, L. Prashad, W. L. Stefanov, and L. Larsen. "Neighborhood Microclimates and Vulnerability to Heat Stress." *Social Science & Medicine* 63, no. 11 (2006): 2847–63.

25. Ibid.

26. E. S. Blake, T. B. Kimberlain, R. J. Berg, J. P. Cangialosi, and J. L. Beven II. "Tropical Cyclone Report: Hurricane Sandy." *National Hurricane Center* 12 (2013): 1–10.

27. Dr. Elmer Pfeffercorn, author (JML), personal reference, https://geiselmed.dartmouth.edu/faculty/facultydb/view.php?uid=1525. Accessed October 2, 2016.

28. WHO, http://www.who.int/csr/don/2012_10_17/en/. Accessed October 4, 2016.

29. S. I. Hay, J. Cox, D. J. Rogers, S. E. Randolph, D. I. Stern, G D. Shanks, . . . and R. W. Snow. "Climate Change and the Resurgence of Malaria in the East African Highlands." *Nature* 415, no. 6874 (2002): 905–9.

30. C. Loiseau, R. J. Harrigan, A. J. Cornel, S. L. Guers, M. Dodge, T. Marzec, . . . and R. N. Sehgal. "First Evidence and Predictions of Plasmodium Transmission in Alaskan Bird Populations." *PLoS One* 7, no. 9 (2012): e44729.

31. C. Bellard, C. Bertelsmeier, P. Leadley, W. Thuiller, and F. Courchamp. "Impacts of Climate Change on the Future of Biodiversity." *Ecology Letters* 15, no. 4 (2012): 365–77.

32. J. Schmidhuber and F. N. Tubiello. "Global Food Security under Climate Change." *Proceedings of the National Academy of Sciences* 104, no. 50 (2007): 19703–8.

33. Ibid.

34. J. Barnett and M. Webber. "Accommodating Migration to Promote Adaption to Climate Change." Policy Research Working Paper 5270, http://poseidon01.ssrn.com/delivery.php?ID=05808407002000911210307612710210201102801502600206002309601900805806104204504700301106909311809609407901706103210708610612209812002009607008712311801712109601609706512712509612707511909402480&EXT=pdf. April 2010. Accessed October 1, 2016.

35. N. J. Grove and A. B. Zwi. "Our Health and Theirs: Forced Migration, Othering, and Public Health." *Social Science & Medicine* 62, no. 8 (2006): 1931–42.

CHAPTER 3

1. J. F. Dhainaut, Y. E. Claessens, C. Ginsburg, and B. Riou. "Unprecedented Heat-Related Deaths during the 2003 Heat Wave in Paris: Consequences on Emergency Departments." *Critical Care* 8, no. 1 (2003): 1.

2. P. A. Stott, D. A. Stone, and M. R. Allen. "Human Contribution to the European Heatwave of 2003." *Nature* 432, no. 7017 (2004): 610–14.

3. N. Christidis, G. S. Jones, and P. A. Stott. "Dramatically Increasing Chance of Extremely Hot Summers since the 2003 European Heatwave." *Nature Climate Change* 5, no. 1 (2015): 46–50.

4. A. Zanobetti, M. S. O'Neill, C. J. Gronlund, and J. D. Schwartz. "Summer Temperature Variability and Long-Term Survival among Elderly People with Chronic Disease." *Proceedings of the National Academy of Sciences* 109, no. 17 (2012): 6608–13.

5. S. L. Harlan, A. J. Brazel, L. Prashad, W. L. Stefanov, and L. Larsen. "Neighborhood Microclimates and Vulnerability to Heat Stress." *Social Science & Medicine* 63, no. 11 (2006): 2847–63.

6. United Nations (UN), http://www.un.org/en/development/desa/news/population/world-urbanization-prospects-2014.html. Accessed October 4, 2016.

7. UN World Population Aging Report, http://www.un.org/en/development/desa/population/publications/pdf/ageing/WPA2015_Report.pdf. Accessed October 4, 2016.

8. National Weather Service, http://www.nws.noaa.gov/om/hazstats.shtml. Accessed October 12, 2016.

CHAPTER 4

1. UN Office for Disaster Risk Reduction (UNISDR). "Human Cost of Weather Related Disasters 1995–2015," https://www.unisdr.org/we/inform/publications/46796. Accessed March 16, 2017.

2. Ibid.

3. Ibid.

4. Centers for Research on the Epidemiology of Disasters, *EM-DAT: The International Disaster Database*, http://www.emdat.be/disaster-trends. Brussels: Ecole de santé publique, Université catholique de Louvain, 2009.

5. B. I. Cook, T. R. Ault, and J. E. Smerdon. "Unprecedented 21st Century Drought Risk in the American Southwest and Central Plains." *Science Advances* 1, no. 1 (2015): e1400082.

6. M. E. Keim. "Extreme Weather Events: The Role of Public Health in Disaster Risk Reduction as a Means for Climate Change Adaption." In *Global Climate Change and Human Health*, edited by G. Luber and J. Lemery, 35–76. San Francisco: John Wiley & Sons, 2015.

7. Centers for Disease Control and Prevention (CDC). "Mortality during a Famine—Gode District, Ethiopia, July 2000." *Morbidity and Mortality Weekly Report* 50, no. 15 (2001): 285–88.

8. F. Assefa, M. Z. Jabarkhil, P. Salama, and P. Spiegel. "Malnutrition and Mortality in Kohistan District, Afghanistan, April 2001." *JAMA* 286, no. 21 (2001): 2723–28.

9. U.S. Army Public Health Command, Charmak Disease, https://phc.amedd.army.mil/PHC%20Resource%20Library/Charmak%20Jan%202010.pdf. Accessed October 5, 2016.

10. K. A. V. R. Krishnamachari, V. Nagarajan, R. Bhat, and T. B. G. Tilak. "Hepatitis Due to Aflatoxicosis: An Outbreak in Western India." *Lancet* 305, no. 7915 (1975): 1061–63.

11. G. Q. Yang, S. Z. Wang, R. H. Zhou, and S. Z. Sun. "Endemic Selenium Intoxication of Humans in China." *American Journal of Clinical Nutrition* 37, no. 5 (1983): 872–81.

12. H. Getahun, A. Mekonnen, R. TekleHaimanot, and F. Lambein. "Epidemic of Neurolathyrism in Ethiopia." *Lancet* 354, no. 9175 (1999): 306–7.

13. N. Mlingi, N. H. Poulter, and H. Rosling. "An Outbreak of Acute Intoxications from Consumption of Insufficiently Processed Cassava in Tanzania." *Nutrition Research* 12, no. 6 (1992): 677–87; M. Ernesto, A. P. Cardoso, D. Nicala, E. Mirione, F. Massaza, J. Cliff, . . . and J. H. Bradbury. "Persistent Konzo and Cyanogen Toxicity from Cassava in Northern Mozambique." *Acta Tropica* 82, no. 3 (2002): 357–62.

14. T. Tylleskär. "The Association between Cassava and the Paralytic Disease Konzo." *Acta Hortic* 375 (1994): 331–40. doi:10.17660/ActaHortic.1994.375.33.

15. J. Kuntz and R. Murray. "Predictability of Swimming Prohibitions by Observational Parameters: A Proactive Public Health Policy, Stamford, Connecticut, 1989–2004." *Journal of Environmental Health* 72, no. 1 (2009): 17–22.

16. M. Bradley, R. Shakespeare, A. Ruwende, M. E. Woolhouse, E. Mason, and A. Munats. "Epidemiological Features of Epidemic Cholera (El Tor) in Zimbabwe." *Transactions of the Royal Society of Tropical Medicine & Hygiene* 90, no. 4 (1996): 378–82.

17. C. Stanke, M. Kerac, C. Prudhomme, J. Medlock, and V. Murray. "Health Effects of Drought: A Systematic Review of the Evidence." *PLOS Currents Disasters*, 2013.

18. B. Cook, R. Miller, and R. Seager. "Did Dust Storms Make the Dust Bowl Drought Worse?" http://www.ldeo.columbia.edu/res/div/ocp/drought/dust_storms.shtml. Lamont-Doherty Earth Observatory, The Earth Institute at Columbia University, 2007.

19. S. R. Gomez, R. A. Parker, J. A. Dosman, and H. H. McDuffie. "Respiratory Health Effects of Alkali Dust in Residents near Desiccated Old Wives Lake." *Archives of Environmental Health* 47, no. 5 (1992): 364–69.

20. CDC. "Increase in coccidioidomycosis—Arizona, 1998–2001." *Morbidity and Mortality Weekly Report* 52, no. 6 (2003): 109–12.

21. J. Chase and T. Knight. "Drought-Induced Mosquito Outbreaks in Wetlands." *Ecology Letters* 6 (2003): 1017–24.

22. B. Edwards, M. Gray, and B. Hunter. "Social and Economic Impacts of Drought on Farm Families and Rural Communities: Submission to the Productivity Commission's Inquiry into Government Drought Support." Melbourne: Australian Institute of Family Studies, 2008.

23. H. Stain, B. Kelly, T. Lewin, N. Higginbotham, J. Beard, and F. Hourihan. "Social Networks and Mental Health among a Farming Population." *Social Psychiatry and Psychiatric Epidemiology* 43 (2008): 843–49.

24. H. Kloos. "Health Aspects of Resettlement in Ethiopia." *Social Science & Medicine* 30, no. 6 (1990): 643–56.

25. National Aeronautics and Space Administration (NASA). "NASA Finds Drought in Eastern Mediterranean Worst of Past 900 Years," http://www.nasa.gov/feature/goddard/2016/nasa-finds-drought-in-eastern-mediterranean-worst-of-past-900-years, March 1, 2016. Accessed October 8, 2016.

26. S. M. Hsiang, M. Burke, and E. Miguel. "Quantifying the Influence of Climate on Human Conflict." *Science* 341, no. 6151 (2013): 1235367.

27. UN Office for Disaster Risk Reduction (UNISDR). "Human Cost of Weather Related Disasters 1995–2015," https://www.unisdr.org/we/inform/publications/46796. Accessed March 16, 2017.

28. Ibid.

29. J. Malilay. "Floods." In *The Public Health Consequences of Disasters*, edited by E. Noji, 287–301. New York: Oxford University Press, 1997.

30. CDC, http://www.cdc.gov/aging/pdf/disaster_planning_goal.pdf. Accessed October 12, 2016.

31. M. Ahern, S. Kovats, P. Wilkinson, R. Few, and F. Matthies. "Global Health Impacts of Floods: Epidemiologic Evidence." *Epidemiologic Reviews* 27 (January 2005): 36–46.

32. Ibid.

33. P. A. Russac. "Epidemiological Surveillance: Malaria Epidemic following the Nino Phenomenon." *Disasters* 10 (1986): 112–17.

34. C. L. Lau, L. D. Smythe, S. B. Craig, and P. Weinstein. "Climate Change, Flooding, Urbanisation and Leptospirosis: Fuelling the Fire?" *Transactions of the Royal Society of Tropical Medicine and Hygiene* 104, no. 10 (2010): 631–38.

35. Ahern, Kovats, Wilkinson, Few, and Matthies. "Global Health Impacts of Floods."

36. Ibid.

37. David M. Theobald, and William H. Romme. "Expansion of the U.S. Wildland—Urban." *Landscape and Urban Planning* 83, no. 4 (2007): 340–54.

38. A. L. Westerling, H. G. Hidalgo, D. R. Cayan, and T. W. Swetnam. "Warming and Earlier Spring Increase Western U.S. Forest Wildfire Activity." *Science* 313, no. 5789 (2006): 940–43.

39. P. L. Kinney. "Climate Change, Air Quality, and Human Health." *American Journal of Preventive Medicine* 35 (2008): 450–67.

40. M. Dennekamp and M. J. Abramson. "The Effects of Bushfire Smoke on Respiratory Health." *Respirology* 16 (2011): 198–209.

41. K. M. Shea, R. T. Truckner, R.W. Weber, and D. B. Peden. "Climate Change and Allergic Disease." *Journal of Allergy and Clinical Immunology* 122 (2008): 443–53.

42. A. G. Rappold, S. L. Stone, W. E. Cascio, L. M. Neas, V. J. Kilaru, M. S. Carraway, . . . and H. Vaughan-Batten. "Peat Bog Wildfire Smoke Exposure in Rural North Carolina Is Associated with Cardiopulmonary Emergency Department Visits Assessed through Syndromic Surveillance." *Environmental Health Perspectives* 119, no. 10 (2011): 1415.

43. J. Schwartz, D. Slater, T. V. Larson, W. E. Pierson, and J. Q. Koenig. "Particulate Air Pollution and Hospital Emergency Room Visits for Asthma in Seattle." *American Review of Respiratory Disease* 147, no. 4 (1993): 826–31.

CHAPTER 5

1. World Health Organization (WHO). "A Global Brief on Vector-Borne Diseases," http://apps.who.int/iris/bitstream/10665/111008/1/WHO_DCO_WHD_2014.1_eng.pdf, 2014. Accessed October 8, 2016. Document number: WHO/DCO/WHD/2014.1.

2. C. D. Beard, J. Garofalo, and K. Gage. "Climate and Its Impacts on Vector-Borne and Zoonotic Disease." In Global Climate Change and Human Health, edited by G. Luber and J. Lemery, 221–66. San Francisco: John Wiley & Sons, 2015.

3. Ibid.

4. WHO. "Malaria," http://www.who.int/mediacentre/factsheets/fs094/en/. Accessed October 4, 2016.

5. Alexandre S. Gagnon, Karen E. Smoyer-Tomic, and Andrew B. Bush. "The El Nino Southern Oscillation and Malaria Epidemics in South America." *International Journal of Biometeorology* 46, no. 2 (2002): 81–89.

6. M. J. Bouma, G. Poveda, W. Rojas, D. Chavasse, M. Quinones, J. Cox, and J. Patz. "Predicting High-Risk Years for Malaria in Colombia Using Parameters of El Niño Southern Oscillation." *Tropical Medicine & International Health* 2, no. 12 (1997): 1122–1127.

7. Kim A. Lindblade, Edward D. Walker, Ambrose W. Onapa, Justus Katungu, and Mark L. Wilson. "Highland Malaria in Uganda: Prospective Analysis of an Epidemic Associated with El Niño." *Transactions of the Royal Society of Tropical Medicine and Hygiene* 93, no. 5 (1999): 480–487.

8. S. I. Hay, J. Cox, D. J. Rogers, S. E. Randolph, D. I. Stern, G. D. Shanks, . . . and R. W. Snow. "Climate Change and the Resurgence of Malaria in the East African Highlands." *Nature* 415, no. 6874 (2002): 905–9.

9. Ibid.

10. A. K. Githeko and W. Ndegwa. "Predicting Malaria Epidemics in the Kenyan Highlands using Climate Data: A Tool for Decision Makers." *Global Change and Human Health* 2, no. 1 (2001): 54–63.

11. K. P. Paaijmans, A. F. Read, and M. B. Thomas. "Understanding the Link between Malaria Risk and Climate." *Proceedings of the National Academy of Sciences* 106, no. 33 (2009): 13844–49.

12. WHO. "Zika," http://www.who.int/mediacentre/factsheets/zika/en/. Accessed October 4, 2016.

13. Ibid.

14. Zika Foundation, https://zikafoundation.org/. Accessed October 4, 2016.

15. Ibid.

16. CDC. "Zika Virus," https://www.cdc.gov/zika/healtheffects/index.html. Accessed October 12, 2016.

17. WHO. "Zika," http://www.who.int/mediacentre/factsheets/zika/en/. Accessed October 4, 2016.

18. Intergovernmental Panel on Climate Change Assessment report 5, chap. 9, Human Health. Working Group II: Impacts, Adaptation and Vulnerability. 9.7.2.1 Modeling the Impact of Climate Change on Dengue, http://www.ipcc.ch/ipccreports/tar/wg2/index.php?idp=361, 2001. Accessed October 8, 2016.

19. Duane J. Gubler. "TMH Dengue, Urbanization and Globalization: The Unholy Trinity of the 21st Century." *Tropical Medicine and Health* 39, no. 4 Supplement (2011): 3–11.

20. CDC. "Dengue," https://www.cdc.gov/dengue/. Accessed October 12, 2016.

21. CDC. "West Nile Virus," https://www.cdc.gov/westnile/. Accessed October 12, 2016.

22. Ibid.

23. Ibid.

24. Ibid.

25. B. J. Johnson and M. V. K. Sukhdeo. "Drought-Induced Amplification of Local and Regional West Nile Virus Infection Rates in New Jersey." *Journal of Medical Entomology* 50, no. 1 (2013): 195–204.

26. M. O. Ruiz, L. F. Chaves, G. L. Hamer, T. Sun, W. M. Brown, E. D. Walker, . . . and U. D. Kitron. "Local Impact of Temperature and Precipitation on West Nile Virus Infection in Culex Species Mosquitoes in Northeast Illinois, USA." *Parasites & Vectors* 3, no. 1 (2010): 1.

27. University of Arizona, UA News. "Effects of Climate Change on West Nile Virus," https://uanews.arizona.edu/story/effects-of-climate-change-on-west-nile-virus, September 9, 2013. Accessed October 8, 2016.

28. CDC. "Lyme Disease," http://www.cdc.gov/lyme/. Accessed October 12, 2016.

29. Ibid.

30. J. S. Brownstein, T. R. Holford, and D. Fish. "Effect of Climate Change on Lyme Disease Risk in North America." *Ecohealth* 2, no. 1 (2005): 38–46.

31. N. H. Ogden, A. Maarouf, I. K. Barker, M. Bigras-Poulin, L. R. Lindsay, M. G. Morshed, . . . and D. F. Charron. "Climate Change and the Potential for Range Expansion of the Lyme Disease Vector Ixodes scapularis in Canada." *International Journal for Parasitology* 36, no. 1 (2006): 63–70.

32. M. Lavelle. "Has Climate Change Made Lyme Disease Worse?" *Scientific American*, https://www.scientificamerican.com/article/has-climate-change-made-lyme-disease-worse/. Accessed October 12, 2016.

33. Beard, Garofalo, and Gage, "Climate and Its Impacts on Vector-Borne and Zoonotic Disease."

CHAPTER 6

1. Intergovernmental Panel on Climate Change. Fifth Assessment Report (AR5), https://www.ipcc.ch/report/ar5/. Accessed October 1, 2016.

2. Regional Health Forum WHO South-East Asia Region 12, no. 1, 2008. "Special Issue on World Health Day 2008 theme: Protecting Health from Climate Change," http://apps.searo.who.int/PDS_DOCS/B3253.pdf. Accessed March 16, 2017.

3. Abdurahman M. El-Sayed and Sandro Galea. "Climate Change and Population Mental Health." In *Global Climate Change and Human Health*, edited by G. Luber and J. Lemery, 317. San Francisco: John Wiley & Sons, 2015.

4. Intergovernmental Panel on Climate Change. Fifth Assessment Report (AR5), https://www.ipcc.ch/report/ar5/. Accessed March 19, 2017.

5. K. J. Aroian and A. E. Norris. "Depression Trajectories in Relatively Recent Immigrants." *Comprehensive Psychiatry* 44, no. 5 (2003): 420–27.

6. R. C. Kessler, S. Galea, and R. T. Jones et al. "Mental Illness and Suicidality after Hurricane Katrina." *Bulletin of the World Health Organization*, http://www.who.int/bulletin/volumes/84/10/06-033019.pdf. doi:10.2471/BLT.06.033019.

7. J. Shukla. "Extreme Weather Events and Mental Health: Tackling the Psychosocial Challenge." *ISRN Public Health*, 2013.

8. R. C. Kessler, S. Galea, and M. J. Gruber et al. "Trends in Mental Illness and Suicidality after Hurricane Katrina." *Mol Psychiatry* 13, no. 4 (2008): 374–84.

9. G. Sullivan, J. J. Vasterling, X. Han, A. T. Tharp, T. Davis, E. A. Deitch, and J. I. Constans. "Preexisting Mental Illness and Risk for Developing a New Disorder after Hurricane Katrina." *Journal of Nervous and Mental Disease* 201 (2013): 161–66. doi:10.1097/NMD.0b013e31827f636d.

10. R. C. Kessler, S. Aguilar-Gaxiola, and J. Alonso et al. "The Global Burden of Mental Disorders: An Update from the WHO World Mental Health (WMH) Surveys." *Epidemiol Psichiatr Soc* 18, no. 1 (2009): 23–33.

11. 2014 National Climate Assessment, http://nca2014.globalchange.gov/highlights/report-findings/extreme-weather. Accessed November 28, 2016.

12. M. Ives and J. Haner."A Remote Pacific Nation, Threatened by Rising Seas." *New York Times*, July 2, 2016, http://www.nytimes.com/2016/07/03/world/asia/climate-change-kiribati.html?_r=0.

13. P. Sykes. "Sinking States; Climate Change and the Next Refugee Crisis," https://www.foreignaffairs.com/articles/2015-09-28/sinking-states, September 28, 2015. Accessed October 8, 2016.

14. CIA World Factbook. "Tuvalu," https://www.cia.gov/library/publications/the-world-factbook/geos/tv.html. Accessed March 16, 2017.

15. P. Sykes. "Sinking States; Climate Change and the Next Refugee Crisis."

16. Ibid.

17. Ibid.

18. Ibid.

19. A. M. El-Sayed, and S. Galea. "Climate Change and Population Mental Health." In *Global Climate Change and Human Health*, edited by G. Luber and J. Lemery, 311–31. San Francisco: John Wiley & Sons, 2015.

20. Steven Hyman et al. "Mental Disorders." *Disease Control Priorities Related to Mental, Neurological, Developmental and Substance Abuse Disorders* (2006): 1–20.

21. C. Mathers, D. M. Fat, and J. Boerma. *The Global Burden of Disease: 2004, Update*. Geneva: World Health Organization, 2008.

22. C. Raleigh. "The Search for Safety: The Effects of Conflict, Poverty and Ecological Influences on Migration in the Developing World." *Global Environmental Change* 21 (2011): S82–93.

CHAPTER 7

1. J. E. McCarthy and R. K. Lattanzio. "Ozone Air Quality Standards: EPA's 2015 Revision." Congressional Research Service, https://fas.org/sgp/crs/misc/R43092.pdf, January 25, 2016. Accessed October 8, 2016.

2. A. M. Fiore, V. Naik, D. V. Spracklen, A. Steiner, N. Unger, M. Prather, . . . and A. Dalsøren. "Global Air Quality and Climate." *Chemical Society Reviews* 41, no. 19 (2012): 6663–83.

3. W. J. Guan, X. Y. Zheng, K. F. Chung, and N. S. Zhong. "Impact of Air Pollution on the Burden of Chronic Respiratory Diseases in China: Time for Urgent Action." *Lancet* 388, no. 10054 (2016): 1939–51.

4. UK Health Alliance on Climate Change, http://www.ukhealthalliance. org/new-report-breath-fresh-air-addressing-climate-change-air-pollution-together-health/. Accessed November 28, 2016.

5. K. Knowlton. "Ozone, Oppressive Air Masses, and Degraded Air Quality." In *Global Climate Change and Human Health.* edited by G. Luber and J. Lemery, 137–70. San Francisco: John Wiley & Sons, 2015.

6. B. J. Lee, B. Kim, and K. Lee. "Air Pollution Exposure and Cardiovascular Disease." *Toxicol Res* 30, no. 2 (2014): 71–75.

7. C. A. Pope III, R. T. Burnett, M. J. Thun, E. E. Calle, and G. D. Thurston. "Lung Cancer, Cardiopulmonary Mortality and Long Term Exposure to Fine Particulate Air Pollution." *JAMA* 287 (2002): 1132–41.

8. A. P. Tai, L. J. Mickley, and D. J. Jacob. "Correlations between Fine Particulate Matter (M 2.5) and Meteorological Variables in the United States: Implications for the Sensitivity of PM 2.5 to Climate Change." *Atmospheric Environment* 44, no. 32 (2010): 3976–84.

9. World Health Organization, Europe: Health risks of particulate matter from long-range transboundary air pollution, http://www.euro.who.int/__data/assets/ pdf_file/0006/78657/E88189.pdf. Accessed March 16, 2017.

10. A. Peters, H. E. Wichmann, T. Tuch, J. Heinrich, and J. Heyder. "Respiratory Effects Are Associated with the Number of Ultrafine Particles." *American Journal of Respiratory and Critical Care Medicine* 155, no. 4 (1997): 1376–83.

11. N. Fann, A. D. Lamson, S. C. Anenberg, K. Wesson, D. Risley, and B. J. Hubbell. "Estimating the National Public Health Burden Associated with Exposure to Ambient PM2. 5 and Ozone." *Risk Analysis* 32, no. 1 (2012): 81–95.

12. J. P. Wisnivesky, S. L. Teitelbaum, A. C. Todd et al. "Persistence of Multiple Illnesses in World Trade Center Rescue and Recovery Workers: A Cohort Study." *Lancet* 378 (2011): 898–905.

13. Ibid.

14. D. J. Jacob and D. A. Winner. "Effect of Climate Change on Air Quality." *Atmospheric Environment* 43, no. 1 (2009): 51–63.

15. N. Grulk and S. Schilling. *Air Pollution and Climate Change*, http://www.fs.fed. us/ccrc/topics/air-pollution.shtml. Washington, DC: U.S. Department of Agriculture, Forest Service, Climate Change Resource Center, 2008.

16. N. A. Molfino, S. C. Wright, I. Katz, S. Tarlo, F. Silverman, P. A. McClean, . . . and M. Raizenne. "Effect of Low Concentrations of Ozone on Inhaled Allergen Responses in Asthmatic Subjects." *Lancet* 338, no. 8761 (1991): 199–203.

17. M. Jerrett, R. T. Burnett, C. A. Pope III, K. Ito, G. Thurston, D. Krewski, . . . and M. Thun. "Long-Term Ozone Exposure and Mortality." *New England Journal of Medicine* 360, no. 11 (2009): 1085–95.

18. World Health Organization. *Air Quality Guidelines: Global Update 2005: Particulate Matter, Ozone, Nitrogen Dioxide, and Sulfur Dioxide.* World Health Organization, 2006.

19. M. L. Bell, S. L. McDermott, J. M. Samet, J. M. Zeger, and F. Dominici. "Ozone and Mortality in 95 U.S. Urban Communities, 1987 to 2000." *JAMA* 292 (2004): 2372–78.

20. P. E. Sheffield, K. Knowlton, J. L. Carr, and P. L. Kinney. "Modeling of Regional Climate Change Effects on Ground-Level Ozone and Childhood Asthma." *American Journal of Preventive Medicine* 41 (2011): 251–57.

21. M. S. Friedman, K. E. Powell, L. Hutwagner, L. M. Graham, . . . and W. G. Teague. "Impact of Changes in Transportation and Commuting Behaviors during the 1996 Summer Olympic Games in Atlanta on Air Quality and Childhood Asthma." *JAMA* 285, no. 7 (2001): 897–905.

22. M. Jerrett, R. T. Burnett, C. A. Pope III, K. Ito, G. Thurston, D. Krewski, Y. Shi et al. "Long-Term Ozone Exposure and Mortality." *New England Journal of Medicine* 360 (2009): 1085–95. doi:10.1056/NEJMoa0803894; N. Fann, A. D. Lamson, S. C. Anenberg, K. Wesson, D. Risley, and B. J. Hubbell. "Estimating the National Public Health Burden Associated with Exposure to Ambient PM2. 5 and Ozone." *Risk Analysis* 32, no. 1 (2012): 81–95.

23. K. Knowlton. "Ozone, Oppressive Air Masses, and Degraded Air Quality." In *Global Climate Change and Human Health*, edited by G. Luber and J. Lemery, 138–70. San Francisco: John Wiley & Sons, 2015.

24. J. D. Berman, N. Fann, J. W. Hollingsworth, K. E. Pinkerton, W. N. Rom, A. M. Szema, P. N. Breysse, R. H. White, and F. C. Curriero. "Health Benefits from Large-Scale Ozone Reduction in the United States." *Environmental Health Perspectives* 120 (2012): 1404–10, http://dx.doi.org/10.1289/ehp.1104851.

25. T. W. Wong, T. S. Lau, T. S. Yu, A. Neller, S. L. Wong, W. Tam, and S. W. Pang. "Air Pollution and Hospital Admissions for Respiratory and Cardiovascular Diseases in Hong Kong." *Occupational and Environmental Medicine* 56, no. 10 (1999): 679–83.

26. J. Schwartz. "Short Term Fluctuations in Air Pollution and Hospital Admissions of the Elderly for Respiratory Disease." *Thorax* 50, no. 5 (1995): 531–38.

27. J. Schwartz. "Lung Function and Chronic Exposure to Air Pollution: A Cross-Sectional Analysis of NHANES II." *Environmental Research* 50, no. 2 (1989): 309–21.

28. J. Sunyer, C. Spix, P. Quenel, A. Ponce-de-Leon, A. Pönka, T. Barumandzadeh, . . . and L. Bisanti. "Urban Air Pollution and Emergency Admissions for Asthma in Four European Cities: The APHEA Project." *Thorax* 52, no. 9 (1997): 760–65.

29. S. K. Liu, S. Cai, Y. Chen et al. "The Effect of Pollutional Haze on Pulmonary Function." *J Thorac Dis* 8, no. 1 (2016): E41–56.

CHAPTER 8

1. United Nations Educational, Scientific and Cultural Organization (UNESCO), Paris. "Securing the Food Supply 2001a," http://webworld.unesco.org/water/wwap/facts_figures/food_supply.shtml. Accessed October 12, 2016.

2. "Climate Change and Water." IPPC Technical Paper VI, http://ipcc.ch/pdf/technical-papers/climate-change-water-en.pdf. Accessed April 10, 2016.

3. Li Liu et al. "Global, Regional, and National Causes of Child Mortality: An Updated Systematic Analysis for 2010 with Time Trends since 2000." *Lancet* 379, no. 9832 (2012): 2151–61.

4. United Nations Water Thematic Paper, "Transboundary Waters: Sharing Benefits, Sharing Responsibilities," http://www.unwater.org/downloads/UNW _TRANSBOUNDARY.pdf, 2008. Accessed October 12, 2016.

5. http://www.cfr.org/global/global-conflict-tracker/p32137#!/conflict/civil-war-in-syria. United Nations High Commissioner for Refugees. "Global Trends; Forced Displacement in 2015," http://www.unhcr.org/576408cd7.pdf, 2015. Accessed October 8, 2016.

6. "Drought in Eastern Mediterranean Worst of Past 900 Years." NASA Global Climate Change, http://climate.nasa.gov/news/2408/, March 1, 2016. Accessed May 6, 2016.

7. Benjamin I. Cook et al. "Spatiotemporal Drought Variability in the Mediterranean Over the Last 900 Years." *Journal of Geophysical Research: Atmospheres* (2016).

8. P. H. Gleick. "The Syrian Conflict and the Role of Water." In *The World's Water*, edited by P. H. Gleick, 147–51. Washington, DC: Island Press/Center for Resource Economics, 2014.

9. International Year of Sanitation 2008: Overview. United Nations Children's Emergency Fund (UNICEF), http://esa.un.org/iys/docs/IYS%20PRESS%20KIT. pdf, 2008. Accessed May 7, 2016.

10. J. T. Watson, M. Gayer, and M. A. Connolly. "Epidemics after Natural Disasters." *Emerg Infect Dis* 13, no. 1 (2007): 1–5.

11. M. Chen, C. Lin, Y. Wu et al. "Effects of Extreme Precipitation to the Distribution of Infectious Diseases in Taiwan, 1994–2008." *PLoS One* 7, no. 6 (2012): e34651.

12. J. Semenza. "Changes in Hydrology and Its Impacts on Waterborne Disease." In *Global Climate Change and Human Health,* edited by G. Luber and J. Lemery, 103–35. San Francisco: John Wiley & Sons, 2015.

13. Ibid.

14. World Health Organization (WHO). "Global Epidemics and Impact of Cholera," http://www.who.int/topics/cholera/impact/en/, 2016. Accessed May 6, 2016.

15. D. R. Karaolis, R. Lan, and P. R. Reeves. "The Sixth and Seventh Cholera Pandemics Are Due to Independent Clones Separately Derived from Environmental, Nontoxigenic, Non-O1 Vibrio Cholera." *J Bacteriolm* 177, no. 11 (June 1995): 3191–98.

16. WHO. "Diarrhea," http://www.who.int/mediacentre/factsheets/fs330/en/. Accessed October 12, 2016.

17. WHO. "Floods in the WHO European Region: Health Effects and Their Preventions," https://www.researchgate.net/publication/275345314_Floods_and_Public_Health_Consequences_Prevention_and_Control_Measures, January 9, 2016. Accessed May 7, 2016.

18. R. H. Davies and C. Wray. "Seasonal Variations in the Isolation of Salmonella typhimurium, Salmonella enteritidis, Bacillus cereus and Clostridium perfringens from

Environmental Samples." *Journal of Veterinary Medicine*, ser. B 43, nos. 1-10 (1996): 119–27.

19. J. B. McLaughlin, A. DePaola, C. A. Bopp, K. A. Martinek, N. P. Napolilli, C. G. Allison, . . . and J. P. Middaugh. "Outbreak of Vibrio parahaemolyticus gastro-enteritis Associated with Alaskan Oysters." *New England Journal of Medicine* 353, no. 14 (2005): 1463–70.

20. V. T. P. Sedas. "Influence of Environmental Factors on the Presence of Vibrio cholerae in the Marine Environment: A Climate Link." *Journal of Infection in Developing Countries* 1, no. 3 (2007): 224–41.

21. R. Philipsborn, S. M. Ahmed, B. J. Brosi, and K. Levy. "Climatic Drivers of Diarrheagenic Escherichia coli Incidence: A Systematic Review and Meta-analysis." *Journal of Infectious Diseases* (2016): jiw081.

22. Ibid.

23. CDC. "Shigella," https://www.cdc.gov/shigella/. Accessed October 12, 2016.

24. S. Kernéis, P. J. Guerin, L. von Seidlein et al. "A Look Back at an Ongoing Problem: Shigella dysenteriae Type 1 Epidemics in Refugee Settings in Central Africa (1993–1995)." Edited by D. J. Diemert. *PLoS ONE* 4, no. 2 (2009): e4494. doi:10.1371/journal.pone.0004494.

25. M. L. Bennish and B. J. Wojtyniak. "Mortality Due to Shigellosis: Community and Hospital Data." *Rev Infect Dis* 13, no. 4 (March–April 1991): S245–51.

26. K. L. Kotloff, J. P. Winickoff, B. Ivanoff et al. "Global Burden of *Shigella* Infections: Implications for Vaccine Development and Implementation of Control Strategies." WHO, http://www.who.int/bulletin/archives/77(8)651.pdf, 1999. Accessed May 6, 2016.

27. CDC. "Typhoid Fever," http://www.cdc.gov/typhoid-fever/. Accessed October 12, 2016.

28. J. A. Crump, S. P. Luby, and E. D. Mintz. "The Global Burden of Typhoid Fever." *Bull World Health Organ* 82 (2004): 346–53.

29. R. C. Dart. *Medical Toxicology*, 3rd ed. Philadelphia: Lippincott Williams & Wilkins, 2004.

30. G. A. Soper. "The Curious Career of Typhoid Mary." *Bulletin of the New York Academy of Medicine* 15, no. 10 (1939): 698.

31. A. Doyle, D. Barataud, A. Gallay et al. "Norovirus Foodborne Outbreaks Associated with the Consumption of Oysters from the Etang de Thau, France, December 2012." *Euro Surveill* 9, no. 3 (March 2004): 24–26.

32. D. Chmid, I. Lederer, P. Much et al. "Outbreak of Norovirus Infection Associated with Contaminated Flood Water, Salzburg 2005." *Eurosurveillance* 10, no. 24 (2005): 2727.

33. E. T. Isakbaeva, M. A. Widdowson, R. S. Beard, S. N. Bulens, J. Mullins, S. S. Monroe, . . . and R. I. Glass. "Norovirus Transmission on Cruise Ship." *Emerg Infect Dis* 11, no. 1 (2005): 154–58.

34. P. C. Carling, L. A. Bruno-Murtha, and J. K. Griffiths. "Cruise Ship Environmental Hygiene and the Risk of Norovirus Infection Outbreaks: An Objective Assessment of 56 Vessels over 3 Years." *Clinical Infectious Diseases* 49, no. 9 (2009): 1312–17.

35. Centers for Disease Control and Prevention (CDC). "Norovirus Outbreak among Evacuees from Hurricane Katrina—Houston, Texas, September 2005," https://www.cdc.gov/mmwr/preview/mmwrhtml/mm5440a3.htm, October 14, 2005. Accessed October 5, 2016.

36. S. Patra, A. Kumar, S. S. Trivedi, M. Puri, and S. K. Sarin. "Maternal and Fetal Outcomes in Pregnant Women with Acute Hepatitis E Virus Infection." *Ann Intern Med* 147 (2007): 28033. Accessed May 6, 2016.

37. Poliomyelitis Pinkbook. CDC, http://www.cdc.gov/vaccines/pubs/pink book/downloads/polio.pdf, April 2015. Accessed May 6, 2016.

38. MMWR Centers for Disease Control and Prevention. "Giardiasis Surveillance—United States, 2011–2012," http://www.cdc.gov/mmwr/preview/mmwrhtml/ss6403a2.htm, May 1, 2015. Accessed May 6, 2016.

39. CDC, *Morbidity and Mortality Weekly Report*. "Cryptosporidiosis Surveillance—United States, 2011–2012," http://www.cdc.gov/mmwr/preview/mmwrhtml/ss6403a1.htm, May 1, 2015. Accessed May 6, 2016.

40. CDC, *Morbidity and Mortality Weekly Report*. M. C. Hlavsa, V. A. Roberst, A. M. Kahler et al. "Outbreaks of Illness Associated with Recreational Water—United States, 2011–2012," http://www.cdc.gov/mmwr/preview/mmwrhtml/mm6424a4.htm, June 26, 2015. Accessed May 6, 2016.

41. R. A. Dillingham, A. A., Lima, and R. L. Guerrant. "Cryptosporidiosis: Epidemiology and Impact." *Microbes and Infection* 4, no. 10 (2002): 1059–66.

42. CDC. "Giardiasis Surveillance—United States, 2011–2012," http://www.cdc.gov/mmwr/preview/mmwrhtml/ss6403a2.htm, May 1, 2015. Accessed May 6, 2016.

43. K. J. Esch and C. A. Petersen. "Transmission and Epidemiology of Zoonotic Protozoal Diseases of Companion Animals." *Clin Microbiol Rev* 26, no. 1 (January 2013): 58–85.

44. WHO. "Soil-Transmitted Helminth Infections Fact Sheet," http://www.who.int/mediacentre/factsheets/fs366/en/, March 2016. Accessed May 6, 2016.

45. N. R. de Silva, S. Brooker, P. J. Hotez, A. Montresor, D. Engles, and L. Savioli. "Soil-Transmitted Helminth Infections: Updating the Global Picture." *Trends in Parasitology* 19 (2003): 547–51.

46. M. J. Chusid, D. C. Dale, B. C. West, and S. M. Wolff. "The Hypereosinophilic Syndrome: Analysis of Fourteen Cases with Review of the Literature." *Medicine* 54, no. 1 (1975): 1–27.

47. M. Kassalik and K. Mönkemüller. "Strongyloides stercoralis Hyperinfection Syndrome and Disseminated Disease." *Gastroenterol Hepatol* 7 (2011): 766–68.

48. Swaytha Ganesh and Ruy J. Cruz Jr. "Strongyloidiasis: A Multifaceted Disease." *Gastroenterol Hepatol (NY)* 7, no. 3 (2011): 194–96.

49. WHO. "Soil-Transmitted Helminth Infections Fact Sheet."

50. K. Annadurai, R. Danasekaran, and G. Mani. "Global Eradication of Guinea Worm Disease: Toward a Newer Milestone." *J Res Med Sci* 19, no. 12 (2014): 1207–8.

51. CDC. "Dracunculiasis," http://www.cdc.gov/parasites/guineaworm/index.html. Accessed October 12, 2016.

52. E. Schwartz. "Schistosomiasis." In *Tropical Diseases in Travelers*, edited by E. Schwartz. San Francisco: John Wiley & Sons, 2009.

53. L. Chitsulo, D. Engels, A. Montresor, and L. Savioli. "The Global Status of Schistosomiasis and Its Control." *Acta Trop* 77 (2000): 41–51. 10.1016/S0001-706X(00)00122-4.

54. WHO. "Schistosomiasis Fact Sheet," http://www.who.int/schistosomiasis/epidemiology/en/. Accessed March 16, 2017.

55. Musa M. Kheir et al. "Mortality Due to Schistosomiasis mansoni: A Field Study in Sudan." *American Journal of Tropical Medicine and Hygiene* 60, no. 2 (1999): 307–10.

56. S. Mas-Coma, M. A. Valero, and M. D. Bargues. "Climate Change Effects on Trematodiases, with Emphasis on Zoonotic Fascioliasis and Schistosomiasis." *Veterinary Parasitology* 163, no. 4 (2009): 264–80.

57. X. N. Zhou, G. J. Yang, K. Yang, X. H. Wang, Q. B. Hong, L. P. Sun, . . . and J. Utzinger. "Potential Impact of Climate Change on Schistosomiasis Transmission in China." *American Journal of Tropical Medicine and Hygiene* 78, no. 2 (2008): 188–94.

58. UN World Water Development Report, http://www.unwater.org/publications/world-water-development-report/en/. Accessed April 10, 2016.

CHAPTER 9

1. Ecosystem Services, Biodiversity and Human Health, Harvard School of Public Health, http://www.chgeharvard.org/topic/ecosystem-services 2012–2016. Accessed June 4, 2016.

2. Daniel Silverstein, "Food Security." Foundations for Global Health Responders Massive Open Online Course (Coursera, 2015), https://www.coursera.org/learn/ghresponder.

3. W. E. Easterling, P. Aggarwal, P. Batima, K. Brander, L. Erda, S. Howeden, A. Kirilenko et al. *Food, Fibre and Forest Products. Climate Change 2007: Impacts, Adaptation and Vulnerability. Contribution of Working Group II to the Fourth Assessment Report of the Intergovernmental Panel on Climate Change*, edited by M. L. Parry, 273–13. Cambridge: Cambridge University Press, 2007.

4. J. R. Porter, L. Xie, A. J. Challinor, K. Cochrane, S. M. Howden, M. M. Iqbal, D. B. Lobell et al. "Food Security and Food Production Systems." In *Climate Change 2014: Impacts, Adaptation, and Vulnerability. Contribution of Working Group II to the Fifth Assessment Report of the Intergovernmental Panel on Climate Change*, edited by C. B. Field, V. R. Barros, D. Dokken, K. J. Mach, M. D. Mastrandrea, T. E. Bilir, M. Chatterjee et al., 485–533. Cambridge: Cambridge University Press, 2014.

5. Daniel R. Taub, Brian Miller, and Holly Allen. "Effects of Elevated CO2 on the Protein Concentration of Food Crops: A Meta-analysis." *Global Change Biology* 14, no. 3 (2008): 565–75.

6. National Oceanic and Atmospheric Administration (NOAA). "Ocean Acidification: The Other Carbon Dioxide Problem," http://www.pmel.noaa.gov/co2/story/Ocean+Acidification. Accessed June 4, 2016.

7. The Australian Government Great Barrier Reef Marine Park Authority. "Impacts of Ocean Acidification on the Reef," http://www.gbrmpa.gov.au/managing

-the-reef/threats-to-the-reef/climate-change/how-climate-change-can-affect-the
-reef/ocean-acidification, 2016. Accessed June 4, 2016.

8. UK Royal Society. *Ocean Acidification Due to Increasing Atmospheric Carbon Dioxide*. Cardiff: Clyvedone Press, 2005.

9. D. Pimentel. "Climate Changes and Food Supply." *Forum for Applied Research and Public Policy* 8, no. 4 (1993): 54–60.

10. E. Callaway. "Pathogen Genome Tracks Irish Potato Famine Back to Its Roots." *Nature* (May 2013). doi:10.1038/nature.2013.13021.

11. E. Barford. "Crop Pests Advancing with Global Warming." *Nature* (September 2013). doi:10.1038/nature.2013.13644.

12. W. R. Cline. *Global Warming and Agriculture: Impact Estimates by Country*. Washington, DC: Center for Global Development and Peterson Institute for international Economics, 2007.

13. Easterling, Aggarwal, Batima, et al. *Food, Fibre and Forest Products*.

14. M. C. Tirado, P. Crahay, L. Mahy et al. "Climate Change and Nutrition: Creating a Climate for Nutrition Security." *Food and Nutrition Bulletin* 34 (2013): 4.

15. D. Clark and A. Garner. "The First 1000 Days: Do YOU See What We See?" *American Academy of Pediatrics*, https://www.aap.org/en-us/advocacy-and-policy/aap-health-initiatives/EBCD/Documents/ALF-1000days.pdf.

16. United Nations Children's Emergency Fund (UNICEF). "First 1,000 Days Last Forever: Scaling Up Nutrition for a Just World," http://www.unicef.org/tajikistan/Op-Ed_in_support_of_UNICEF_global_nutrition_report_adopted_TJK_facts_ENG.pdf, 2012. Accessed June 4, 2016.

17. Clark and Garner, "The First 1000 Days."

18. Silverstein, "Food Security."

19. M. A. Fernando. "Effect of Ascaris lumbricoides Infestation on Growth of Children." *Indian Pediatrics* 20, no. 10 (1983): 721.

20. L. Stephenson. *The Impact of Helminth Infections on Human Nutrition*. London: Taylor and Francis, 1987.

21. S. Mas-Coma, A. M. Valero, and M. D. Bargues. "Climate Change Effects on Trematodiases, with Emphasis on Zoonotic Fascioliasis and Schistosomiasis." *Veterinary Parasitology* 163, no. 4 (August 2009): 264–80.

22. K. Watkins. Human Development Report, UN Development Programme, 2007.

23. S. J. Lloyd, R. S. Kovats, and Z. Chalabi. "Climate Change, Crop Yields, and Undernutrition: Development of a Model to Quantify the Impact of Climate Scenarios on Child Undernutrition." *Environmental Health Perspectives* 119 (2011): 1817.

24. G. W. Yohe, R. D. Lasco, Q. K. Ahmand, N. W. Arnell, S. J. Cohen, C. Hope et al. "Perspectives on Climate Change and Sustainability." In *Climate Change 2007: Impacts, Adaptation and Vulnerability, Contribution of Working Group II to the Fourth Assessment Report of the Intergovernmental Panel on Climate Change*, edited by M. L. Parry, O. F. Canziani, J. P. Palutikof, P. J. van der Linden, and C. E. Hanson, 811–41. Cambridge: Cambridge University Press, 2007.

25. Maria Cristina Tirado et al. "Climate Change and Nutrition: Creating a Climate for Nutrition Security." *Food and Nutrition Bulletin* 34, no. 4 (2013): 53–47.

26. WHO. "Taking Action, Nutrition for Survival, Growth & Development," http://www.who.int/pmnch/topics/child/acf_whitepaper.pdf, May 2010. Accessed June 4, 2016.

27. Silverstein, "Food Security."

28. S. B. Idso and K. E. Idso. "Effects of Atmospheric Carbon Dioxide Enrichment on Plant Constituents Related to Animal and Human Health." *Environmental and Experimental Botany* 45 (2001): 179–99.

29. I. Loladze. "Hidden Shift of the Inome of Plants Exposed to Elevated CO2 Depletes Minerals at the Base of Human Nutrition." *eLife* 3 (2014): e02245. doi:107554/Elife.02245.

30. Silverstein, "Food Security."

31. "Healthy Homes Issues: Pesticides in the Home—Use, Hazards, and Integrated Pest Management," http://portal.hud.gov/hudportal/documents/huddoc?id=DOC_12484.pdf, March 2006. Accessed June 4, 2016.

32. C. A. Damalas and I. G. Eleftherohorinos. "Pesticide Exposure, Safety Issues, and Risk Assessment Indicators." *Int J Environ Res Public Health* 8, no. 5 (May 2011): 1402–19. doi:10.3390/ijerph8051402.

33. Damalas and Eleftherohorinos, "Pesticide Exposure, Safety Issues, and Risk Assessment Indicators."

34. L. Field, J. W. Kern, and L. B. Rosman. "Re-visiting Projections of PCBs in the Lower Hudson River Fish Using Model Emulation." *Science of the Total Environment* (July 2016): 557–58, 489–501. doi:10.1016/j.scitotenv.2016.02.072.

35. Centers for Disease Control and Prevention (CDC), Agency for Toxic Substance and Disease Registry. "Public Health Statement for DDT, DDE, and DDD," http://www.atsdr.cdc.gov/phs/phs.asp?id=79&tid=20, September 2002. Accessed June 4, 2016.

36. G. D. Veith. "Baseline Concentrations of Polychlorinated Biphenyls and DDT in Lake Michigan Fish, 1971." *Pestic Monit J* 9, no. 1 (June 1975): 21–29.

37. Lake Michigan Management Plan 2000, https://www.epa.gov/sites/production/files/2015-11/documents/lake-michigan-lamp-2000-458pp.pdf, April 2000. Accessed June 4, 2016.

38. B. C. Kelly, M. G. Ikonomou, J. D. Blair, A. E. Morin and F. A. Gobas. Food web–specific biomagnification of persistent organic pollutants. *Science*, 317 no. 5835 (2007): 236–39.

39. United States Environmental Protection Agency: https://www.epa.gov/international-cooperation/persistent-organic-pollutants-global-issue-global-response#resources. Accessed June 15, 2017.

40. Arctic Climate Impact Assessment (ACIA). *Impacts of a Warming Arctic: Arctic Climate Impact Assessment* (ACIA Overview Report). Cambridge: Cambridge University Press, 2004.

41. S. Booth and D. Zeller. "Mercury, Food Webs, and Marine Mammals: Implications of Diet and Climate Change for Human Health," *Environmental Health Perspectives* 113 (2005): 521–26.

42. EPA, https://www.epa.gov/mercury/how-people-are-exposed-mercury. Accessed November 28, 2016.

43. J. L. DesGranges, J. Rodrigue, B. Tardif, and M. Laperle. "Mercury Accumulation and Biomagnification in Ospreys (Pandion haliaetus) in the James Bay and Hudson Bay Regions of Quebec." *Archives of Environmental Contamination and Toxicology* 35, no. 2 (1998): 330–41.

44. P. Muriel, T. Downing, M. Hulme, R. Harrington, D. Lawlor, D. Wurr, C. J. Atkinson et al. *Climate Change and Agriculture in the United Kingdom*. London: Ministry of Agriculture, Fisheries and Forestry, 2001.

45. S. W. Baily. "Climate Change and Decreasing Herbicide Persistence." *Pest Management Science* 60 (2004): 158–62.

46. M. D. Awasthi, A. K. Ahuja, and D. Sharma. "Contamination of Horticulture Ecosystem: Orchard Soil and Water Bodies with Pesticide Residues." Proceedings of National Symposium on Integrated Pest Management (IPM) in Horticultural Crops: New Molecules. Biopesticides and Environment, 2001.

47. B. Gevao, K. T. Semple, and K. C. Jones. "Bound Pesticide Residues in Soils: A Review." *Environmental Pollution* 108, no. 1 (2000): 3–14.

48. J. R. Roberts, C. J. Karr, J. A. Paulson, A. C. Brock-Utne, H. L. Brumberg, C. C. Campbell, . . . and R. O. Wright. "Pesticide Exposure in Children." *Pediatrics* 130, no. 6 (2012): e1765–88.

49. J. Jeyaratnam. "Health Problems of Pesticide Usage in the Third World." *British Journal of Industrial Medicine* 42 (1985a): 505–6.

50. R. Spiewak. "Pesticides as a Cause of Occupational Skin Diseases in Farmers." *Ann Agric Environ Med* 8, no. 1 (2001): 1–5.

51. M. P. Montgomery, F. Kamel, T. M. Salana, M. C. R. Alavanja, and D. P. Sandler. "Incident Diabetes and Pesticide Exposure among Licensed Pesticide Applicators: Agricultural Health Study 1993–2003." *Amer J Epidemiol* 167 (2008): 1235.

52. R. M. Salvi, D. R. Lara, E. S. Ghisolfi, L. V. Portela, R. D. Dias, and D. O. Souza. "Neuropsychiatric Evaluation in Subjects Chronically Exposed to Organophosphate Pesticides." *Toxicological Sciences* 72, no. 2 (2003): 267–71.

53. S. H. Zahm and A. Blair. "Pesticides and Non-Hodgkin's Lymphoma." *Cancer Res* 52, 19 suppl (October 1992): 5485–88.

54. M. Moses. "Pesticide-Related Health Problems and Farmworkers." *Official Journal of the American Association of Occupational Health Nurses* 37, no. 3 (1989): 115–30.

55. Ibid.

56. M. Edwards, D. G. Johns, S. C. Leterme, E. Svendsen, and A. J. Richardson. "Regional Climate Change and Harmful Algal Blooms in the Northeast Atlantic." *Limnology and Oceanography* 51 (2006): 820–29.

57. K. G. Sellner, G. J. Doucette, and G. J. Kirkpatric. "Harmful Algal Blooms: Causes, Impacts and Detection." *Journal of Industrial Microbiology and Biotechnology* 30 (2003): 383–406.

58. CDC. "Pfiesteria," http://www.cdc.gov/hab/pfiesteria/pdfs/about.pdf. Accessed October 12, 2016.

59. Natural Resources Defense Council. M. Dorfman, N. Stoner, and M. Merkel. "Swimming in Sewage," https://www.nrdc.org/sites/default/files/sewage.pdf, February 2004. Accessed June 9, 2016.

60. EPA. "Health Risks of Human Exposure to Wastewater," https://nepis.epa.gov/
Exe/ZyNET.exe/20016SRO.TXT?ZyActionD=ZyDocument&Client=EPA&Index
=1981+Thru+1985&Docs=&Query=&Time=&EndTime=&SearchMethod=1&Toc
Restrict=n&Toc=&TocEntry=&QField=&QFieldYear=&QFieldMonth=&QFieldD
ay=&IntQFieldOp=0&ExtQFieldOp=0&XmlQuery=&File=D%3A%5Czyfiles%5CIn
dex%20Data%5C81thru85%5CTxt%5C00000013%5C20016SRO.txt&User=ANONY
MOUS&Password=anonymous&SortMethod=h%7C-&MaximumDocuments=1&Fuz
zyDegree=0&ImageQuality=r75g8/r75g8/x150y150g16/i425&Display=hpfr&DefSeek
Page=x&SearchBack=ZyActionL&Back=ZyActionS&BackDesc=Results%20page&M
aximumPages=1&ZyEntry=1&SeekPage=x&ZyPURL. Accessed November 28, 2016.

61. EPA. "Murphy Oil Spill Factsheet," http://www.columbia.edu/itc/journalism/
cases/katrina/Federal%20Government/Environmental%20Protection%20Agency/Mur
phy%20Oil%20Spill%20Fact%20Sheet%20Feb%202006.pdf, February 2006. Accessed
October 12, 2016.

62. G. Umlauf, G., Bidoglio, E. H. Christoph, J. Kampheus, F. Krüger, D. Landmann, . . . and D. Stehr. "The Situation of PCDD/Fs and Dioxin-like PCBs after
the Flooding of River Elbe and Mulde in 2002." *Acta hydrochimica et hydrobiologica* 33,
no. 5 (2005): 543–54.

CHAPTER 10

1. L. H. Ziska and K. L. Ebi. "Climate Change, Carbon Dioxide, and Public
Health." In *Global Climate Change and Human Health*, edited by G. Luber and J. Lemery, 195–213. San Francisco: John Wiley & Sons, 2015.

2. Colleen E. Reid and Janet L. Gamble. "Aeroallergens, Allergic Disease, and Climate Change: Impacts and Adaptation." *Ecohealth* 6, no. 3 (2009): 458–70.

3. Ziska and Ebi, "Climate Change, Carbon Dioxide, and Public Health."

4. Ibid.

5. E. J. O'Connell. "The Burden of Atopy and Asthma in Children." *Allergy* 59,
suppl 78 (2004): 7–11.

6. Ziska and Ebi, "Climate Change, Carbon Dioxide, and Public Health."

7. Ibid.

8. R. W. Weber. "Patterns of Pollen Cross-Allergenicity." *Journal of Allergy and
Clinical Immunology* 112, no. 2 (2003): 229–39.

9. E. Lo and E. Levetin. "Influence of Meteorological Conditions on Early Spring
Pollen in the Tulsa Atmosphere from 1987–2006." *Journal of Allergy and Clinical Immunology* 119, no. 1 (2007): S101.

10. H. B. Freye, J. King, and C. M. Litwin. "Variations of Pollen and Mold Concentrations in 1998 during the Strong El Nino Event of 1997–1998 and Their Impact
on Clinical Exacerbations of Allergic Rhinitis, Asthma, and Sinusitis." *Allergy and
Asthma Proceedings* 22 (2001): 239–47.

11. L. H. Ziska, K. Knowlton, C. Rogers, D. Dalan, N. Tierney, M. A. Elder, W.
Filley et al. "Recent Warming by Latitude Associated with Increased Length of Ragweed

Pollen Season in Central North America." *Proceedings of the National Academy of Sciences USA* 108 (2001): 4248–51.

12. A. Celenza, J. Fothergill, E. Kupek, and R. J. Shaw. "Thunderstorm Associated Asthma: A Detailed Analysis of Environmental Factors." *BMJ* 312, no. 7031 (1996): 604–7.

13. American College of Allergy, Asthma & Immunology. "The Year 2040: Double the Pollen, Double the Allergy Suffering," http://acaai.org/news/year-2040-double-pollen-double-allergy-suffering, 2014. Accessed September 26, 2016.

14. R. Ratard, C. M. Brown, J. Ferdinands, and D. Callahan. "Health Concerns Associated with Mold in Water-Damaged Homes after Hurricanes Katrina and Rita—New Orleans, Louisiana, October 2005." *Morb Mortal Wkly Rep* 55, no. 2 (March 10, 2006): 41–44.

15. J. Manuel. "In Katrina's Wake." *Environmental Health Perspective* 114, no. 1 (2006): A32–39.

16. Paul R. Epstein and Evan Mills. "Climate Change Futures: Health, Ecological and Economic Dimensions." Center for Health and the Global Environment, Harvard Medical School, 2005.

17. CDC. "Coccidioidomycosis," http://www.cdc.gov/fungal/diseases/coccidioidomycosis/. Accessed November 28, 2016.

18. D. Griffin and K. J. Anchukaitis. "How Unusual Is the 2012–2014 California Drought?" Geophysical Research Letters. American Geophysical Union, 2014. doi:10.1002/2014GL062433.

19. G. D'Amato, G. Liccardi, M. D'Amato, and M. Cazzola. "Outdoor Air Pollution, Climatic Changes and Allergic Bronchial Asthma." *European Respiratory Journal* 20, no. 3 (2002): 763–76.

20. U.S. Global Change Research Program. "Review of the Impacts of Climate Variability and Change on Aeroallergens and Their Associated Effects," http://static1.1.sqspcdn.com/static/f/551504/6467325/1270769757893/GCRP.pdf?token=w6WDUSqpLQ6D%2B%2BljSLqMKAWExYQ%3D. Accessed March 16, 2017.

21. T. L. Tanner. "Rhus (Toxicodendron) Dermatitis." *Primary Care* 27, no. 2 (2000): 493–502.

22. Ziska and Ebi, "Climate Change, Carbon Dioxide, and Public Health."

23. Ibid.

CHAPTER 11

1. L. C. Backer. "Effects of Climate Change on Noninfectious Waterborne Threats." In *Global Climate Change and Human Health*, edited by G. Luber and J. Lemery, 171–93. San Francisco: John Wiley & Sons, 2015.

2. Center for Earth and Environmental Science. "What Causes Algal Bloom?" http://www.cees.iupui.edu/research/algaltoxicology/bloomfactors. Accessed October 12, 2016.

3. S. Hales, P. Weinstein, and A. Woodward. "Ciguatera (Fish Poisoning), El Niño, and Pacific Sea Surface Temperatures." *Ecosystem Health* 5, no. 1 (1999): 20–25.

4. S. K. Moore, N. J. Mantua, B. M. Hickey, and V. L. Trainer. "Recent Trends in Paralytic Shellfish Toxins in Puget Sound, Relationships to Climate, and Capacity for Prediction of Toxic Events." *Harmful Algae* 8 (2009): 463–77.

5. E. S. Reichwaldt and A. Ghadouani. "Effects of Rainfall Patterns on Toxic Cyanobacterial Blooms in a Changing Climate: Between Simplistic Scenarios and Complex Dynamics." *Water Research* 46, no. 5 (2012): 1372–93.

6. Ibid.

7. D. M. Anderson, J. M. Burkholder, W. P. Cochlan et al. "Harmful Algal Blooms and Eutrophication: Examining Linkages from Selected Coastal Regions of the United States." *Harmful Algae* 8, no. 1 (2008): 39–53.

8. National Centers for Coastal Ocean Science. "How Climate Change Could Impact Harmful Algal Blooms," https://coastalscience.noaa.gov/news/climate/climate-change-impact-harmful-algal-blooms/, November 6, 2014. Accessed October 5, 2016.

9. O. Hoegh-Guldberg, P. J. Mumby, A. J. Hooten, R. S. Steneck, P. Greenfield, E. Gomez, . . . and N. Knowlton. "Coral Reefs under Rapid Climate Change and Ocean Acidification." *Science* 318, no. 5857 (2007): 1737–42.

10. National Centers for Coastal Ocean Science, "How Climate Change Could Impact Harmful Algal Blooms."

11. I. Falconer. "Measurement of Toxins from Blue-Green Algae in Water and Foodstuffs." In *Algal Toxins in Seafood and Drinking Water*, edited by I. R. Falconer, 165–75. Orlando, FL: Academic Press, 1993.

12. I. Falconer. "Algal Toxins and Human Health." In *The Handbook of Environmental Chemistry*, vol. 5, edited by J. Hrub, 53–82. Berlin: Springer-Verlag, 1998.

13. United States Geological Survey, Toxic Substances Hydrology Program, http://toxics.usgs.gov/highlights/algal_toxins/algal_faq.html. Accessed October 12, 2016.

14. S. M. Azevedo, W. W. Carmichael, E. M. Jochimsen, K. L. Rinehart, S. Lau, G. R. Shaw, and G. K. Eaglesham. "Human Intoxication by Microcystins during Renal Dialysis Treatment in Caruaru—Brazil." *Toxicology* 181 (2002): 441–46.

15. Centers for Disease Control and Prevention (CDC). "Haff Disease Associated with Eating Buffalo Fish—United States, 1997." *Morbidity and Mortality Weekly Report* 47, no. 50 (1998): 1091–93.

16. National Ocean Service. "What Is a Red Tide?" http://oceanservice.noaa.gov/facts/redtide.html. Accessed October 5, 2016.

17. L. E. Fleming, D. Kirkpatrick, L. Backer, J. Bean, A. Wanner, D. Dalpra, R. Tamer et al. "Initial Evaluation of the Effects of Aerosolized Florida Red Tide Toxins (Brevetoxins) in Persons with Asthma." *Environmental Health Perspectives* 113: 650–57.

18. B. Kirkpatric, L. E. Fleming, L. C. Backer, J. A. Bean, R. Tamer, G. Kirkpatrick, T. Kane et al. "Environmental Exposures to Florida Red Tides: Effects on Emergency Room Respiratory Diagnoses Admissions." *Harmful Algae* 5 (2006): 526–33.

19. Department of Agriculture Foreign Agricultural Service GAIN Report, http://gain.fas.usda.gov/Recent%20GAIN%20Publications/Red%20Tide%20and%20Labor%20Unrest%20Reduce%20Chilean%20Salmon%20Production_Santiago_Chile_7-5-2016.pdf, July 5, 2016. Accessed November 14, 2016.

20. D. Baden, L. Fleming, and J. Bean. "Marine Toxins." In *Handbook of Clinical Neurology: Intoxications of the Nervous System Part II. Natural Toxins and Drugs*, edited by F. A. deWolff, 141–75. Amsterdam: Elsevier Press, 1995.

21. B. Lazensky. "Florida Department of Health, Nassau County Health Department, Investigation of a Cluster of Ciguatera Fish Poisoning Cases (n = 13) in Restaurant Patrons Who Consumed Grouper." Unpublished data, 2008.

22. H. Fukiki, M. Suganuma, H. Suguri et al. "Diarrhetic Shellfish Toxin, Dinophysistoxin-1, Is a Potent Tumor Promoter on Mouse Skin." *Cancer Science* 79, no. 10 (1988): 1089–93.

23. Ibid.

24. T. M. Perl, L. Bédard, T. Kosatsky, J. C. Hockin, E. C. Todd, and R. S. Remis. "An Outbreak of Toxic Encephalopathy Caused by Eating Mussels Contaminated with Domoic Acid." *New England Journal of Medicine* 322, no. 25 (1990): 1775–80.

25. Ibid.

26. B. Jeffery, T. Barlow, K. Moizer, S. Paul, and C. Boyle. "Amnesic Shellfish Poison." *Food and Chemical Toxicology* 42, no. 4 (2004): 545–57.

27. K. S. Grant, T. M. Burbachker, E. M. Faustman, and L. Grattan. "Domoic Acid: Neurobehavioral Consequences of Exposure to a Prevalent Marine Biotoxin." *Neurotoxicol Teratol* 32, no. 2 (2010): 132–41.

28. D. C. Rodrigue, R. A. Etzel, S. Hall, E. De Porras, O. H. Velasquez, R. V. Tauze, . . . and P. A. Blake. "Lethal Paralytic Shellfish Poisoning in Guatemala." *American Journal of Tropical Medicine and Hygiene* 42, no. 3 (1990): 267–71.

29. T. A. Caller, J. W. Doolin, J. F. Haney et al. "A Cluster of Amyotrophic Lateral Sclerosis in New Hampshire: A Possible Role for Toxic Cyanobacteria Blooms." *Amyotrophic Lateral Sclerosis* 10, no. 2 (2009): 101–8.

30. J. Pablo, S. A. Banack, P. A. Cox, T. E. Johnson, S. Papapetropoulos, W. G. Bradley, . . . and D. C. Mash. "Cyanobacterial Neurotoxin BMAA in ALS and Alzheimer's Disease." *Acta Neurologica Scandinavica* 120, no. 4 (2009): 216–25.

31. P. A. Cox, R. Richer, J. S. Metcalf, S. A. Banack, G. A. Codd, and W. G. Bradley. "Cyanobacteria and BMAA Exposure from Desert Dust: A Possible Link to Sporadic ALS among Gulf War Veterans." *Amyotrophic Lateral Sclerosis* 10, supp 2 (2009): 109–17.

32. R. Richer, S. A. Banack, J. S. Metcalf, and P. A. Cox. "The Persistence of Cyanobacterial Toxins in Desert Soils." *Journal of Arid Environments* 112, no. B (2015): 134–39.

CHAPTER 12

1. E. Chivian and A. Bernstein. *Sustaining Life.* Oxford: Oxford University Press, 2008, 117.

2. E. Chivian and A. Bernstein. Center for Health and the Global Environment, Harvard Medical School. *How Our Health Depends on Biodiversity*, http://www.chge.harvard.org/sites/default/files/resources/182945%20HMS%20Biodiversity%20booklet.pdf. Accessed October 8, 2016.

3. J. Baillie, C. Hilton-Taylor, and S. N. Stuart. *2004 IUCN Red List of Threatened Species: A Global Species Assessment.* International Union for Conservation of Nature and Natural Resources (2004).

4. Bradley C. Bennett. "Doctrine of Signatures: An Explanation of Medicinal Plant Discovery or Dissemination of Knowledge?" *Economic Botany* 61, no. 3 (2007): 246–55.

5. Edmund Stone. "An Account of the Success of the Bark of the Willow in the Cure of Agues. In a Letter to the Right Honourable George Earl of Macclesfield, President of RS from the Rev. Mr. Edmund Stone, of Chipping-Norton in Oxfordshire." *Philosophical Transactions* 53 (1763): 195–200.

6. S. Squires. "The World's Most Popular Pill Turns 100." *Washington Post*, August 5, 1997. https://www.washingtonpost.com/archive/lifestyle/wellness/1997/08/05/aspirin-the-worlds-most-popular-pill-turns-100/caa961d1-c7a4-42c7-b1ac-550193a9a21f/. Accessed October 12, 2016.

7. R. B. Turner and R. B. Woodward. "The Chemistry of the Cinchona Alkaloids." *Alkaloids: Chemistry and Physiology* 3 (1953): 1–63.

8. Ibid.

9. G. M. Cragg and D. J. Newman. "Natural Products: A Continuing Source of Novel Drug Leads." *Biochimica et Biophysica Acta (BBA)—General Subjects* 1830, no. 6 (2013): 3670–95.

10. D. McNeil. "For Intrigue, Malaria Drug Gets the Prize." *New York Times*, January 16, 2012. http://www.nytimes.com/2012/01/17/health/for-intrigue-malaria-drug-artemisinin-gets-the-prize.html. Accessed November 28, 2016.

11. E. Hsu. "Reflections on the 'Discovery' of the Antimalarial Qinghao." *British Journal of Clinical Pharmacology* 61, no. 6 (2006): 666–70.

12. T. Efferth, A. Benakis, M. R. Romero, M. Tomicic, R. Rauh, D. Steinbach, . . . and M. Marschall. "Enhancement of Cytotoxicity of Artemisinins toward Cancer Cells by Ferrous Iron." *Free Radical Biology and Medicine* 37, no. 7 (2004): 998–1009.

13. Ibid.

14. "The State of Clinical Research in the United States: An Overview." In *Transforming Clinical Research in the United States: Challenges and Opportunities: Workshop Summary*. Institute of Medicine (U.S.) Forum on Drug Discovery, Development, and Translation. Washington, DC: National Academies Press, 2010.

15. A. D. Barnosky, N. Matzke, S. Tomiya, G. O. Wogan, B. Swartz, T. B. Quental, . . . and B. Mersey. "Has the Earth's Sixth Mass Extinction Already Arrived?" *Nature* 471, no. 7336 (2011): 51–57.

16. G. Ceballos, P. R. Ehrlich, A. D. Barnosky, A. García, R. M. Pringle, and T. M. Palmer. "Accelerated Modern Human–Induced Species Losses: Entering the Sixth Mass Extinction." *Science Advances* 1, no. 5 (2015): e1400253.

CHAPTER 13

1. R. B. Fuller and J. Snyder. *Operating Manual for Spaceship Earth*. Carbondale: Southern Illinois University Press, 1969.

2. S. G. Potts, J. C. Biesmeijer, C. Kremen, P. Neumann, O. Schweiger, and W. E. Kunin. "Global Pollinator Declines: Trends, Impacts and Drivers." *Trends in Ecology & Evolution* 25, no. 6 (2010): 345–53.

3. J. Melilo and O. Sala. "Ecosystem Services." In *Sustaining Life: How Human Health Depends on Biodiversity.* New York: Oxford University Press, 2008.

4. A. Wellburn. "Atmospheric Nitrogenous Compounds and Ozone—Is NOx Fixation by Plants a Possible Solution?" *New Phytologist* 138, no. 1 (1998): 5–9.

5. Melilo and Sala. "Ecosystem Services."

6. Ibid.

7. K. Fitzpatrick and M. LaGory. *Unhealthy Places: The Ecology of Risk in the Urban Landscape.* New York: Routledge, 2002.

8. G. W. Evans. "The Built Environment and Mental Health." *Journal of Urban Health* 80, no. 4 (2003): 536–55.

9. M. L. Imhoff, L. Bounoua, T. Ricketts, C. Loucks, R. Harriss, and W. T. Lawrence. "Global Patterns in Human Consumption of Net Primary Production." *Nature* 429, no. 6994 (2004): 870–73.

10. S. Rojstaczer, M. Sterling, and N. J. Moore. "Human Appropriation of Photosynthesis Products." *Science* 294, no. 5551 (2001): 2549–52.

11. Intergovernmental Panel on Climate Change. Fifth Assessment Report (AR5), https://www.ipcc.ch/report/ar5/. Accessed October 5, 2016.

12. D. P. C. Peters, A. E. Lugo, F. S. Chapin III, S. T. A. Pickett, M. Duniway, A. V. Rocha, F. J. Swanson, C. Laney, and J. Jones. "Cross-System Comparisons Elucidate Disturbance Complexities and Generalities." *Ecosphere* 2 (2011): 1–26. doi:10.1890/ES11-00115.1.

13. Intergovernmental Panel on Climate Change. Fifth Assessment Report (AR5).

14. C. Shepard, V. N. Agostini, B. Gilmer, T. Allen, J. Stone, W. Brooks, and M. W. Beck. "Assessing Future Risk: Quantifying the Effects of Sea Level Rise on Storm Surge Risk for the Southern Shores of Long Island, New York." *Natural Hazards* 60 (2012): 727–45. doi:10.1007/s11069-011-0046-8.

15. Gordon McGranahan, Deborah Balk, and Bridget Anderson. "The Rising Tide: Assessing the Risks of Climate Change and Human Settlements in Low Elevation Coastal Zones." *Environment and Urbanization* 19, no. 1 (2007): 17–37.

16. S. B. Roy, L. Chen, E. H. Girvetz, E. P. Maurer, W. B. Mills, and T. M. Grieb. "Projecting Water Withdrawal and Supply for Future Decades in the U.S. under Climate Change Scenarios." *Environmental Science & Technology* 46 (2012): 2545–56. doi:10.1021/es2030774.

17. P. A. Raymond, M. B. David, and J. E. Saiers. "The Impact of Fertilization and Hydrology on Nitrate Fluxes from Mississippi Watersheds." *Current Opinion in Environmental Sustainability* 4 (2012): 212–18. doi:10.1016/j.cosust.2012.04.001.

18. P. A., Raymond, N.-H. Oh, E. R. Turner, and W. Broussard. "Anthropogenically Enhanced Fluxes of Water and Carbon from the Mississippi River." *Nature* 451 (2008): 449–52. doi:10.1038/nature06505.

19. "Wildland Fire Summary and Statistics Annual Report 2011." Boise, ID: National Interagency Fire Center, 2012.

20. Roy, Chen, Girvetz, Maurer, Mills, and Grieb. "Projecting Water Withdrawal and Supply for Future Decades in the U.S. under Climate Change Scenarios."

21. J. L. Sabo, T. Sinha, L. C. Bowling, G. H. W. Schoups, W. W. Wallender, M. E. Campana, K. A. Cherkauer, P. L. Fuller, W. L. Graf, J. W. Hopmans, J. S.

Kominoski, C. Taylor, S. W. Trimble, R. H. Webb, and E. E. Wohl. "Reclaiming Freshwater Sustainability in the Cadillac Desert." *Proceedings of the National Academy of Sciences* 107 (2010): 21263–69. doi:10.1073/pnas.1009734108.

22. T. P. Barnett, D. W. Pierce, H. G. Hidalgo, C. Bonfils, B. D. Santer, T. Das, G. Bala, A. W. Wood, T. Nozawa, A. A. Mirin, D. R. Cayan, and M. D. Dettinger. "Human-Induced Changes in the Hydrology of the Western United States." *Science* 319 (2008): 1080–83. doi:10.1126/science.1152538.

23. S. J. Wenger, D. J. Isaak, C. H. Luce, H. M. Neville, K. D. Fausch, J. B. Dunham, D. C. Dauwalter, M. K. Young, M. M. Elsner, B. E. Rieman, A. F. Hamlet, and J. E. Williams. "Flow Regime, Temperature, and Biotic Interactions Drive Differential Declines of Trout Species under Climate Change." *Proceedings of the National Academy of Sciences* 108 (2011): 14175–80. doi:10.1073/pnas.1103097108.

24. P. S. A. Beck, G. P. Juday, C. Alix, V. A. Barber, S. E. Winslow, E. E. Sousa, P. Heiser, J. D. Herriges, and S. J. Goetz. "Changes in Forest Productivity across Alaska Consistent with Biome Shift." *Ecology Letters* 14 (2011): 373–79. doi:10.1111/j.1461-0248.2011.01598.x.

25. A. E. Kelly and M. L. Goulden. "Rapid Shifts in Plant Distribution with Recent Climate Change." *Proceedings of the National Academy of Sciences* 105 (2008): 11823–26. doi:10.1073/pnas.0802891105.

26. Wenger, Isaak, Luce et al. "Flow Regime, Temperature, and Biotic Interactions Drive Differential Declines of Trout Species under Climate Change."

27. J. S. Collie, A. D. Wood, and H. P. Jeffries. "Long-Term Shifts in the Species Composition of a Coastal Fish Community." *Canadian Journal of Fisheries and Aquatic Sciences* 65 (2008): 1352–65. doi:10.1139/F08-048.

28. J. S. Dukes, N. R. Chiariello, S. R. Loarie, and C. B. Field. "Strong Response of an Invasive Plant Species (Centaurea solstitialis L.) to Global Environmental Changes." *Ecological Applications* 21 (2011): 1887–94. doi:10.1890/11-0111.1.

29. A. J. Eagle, M. E. Eiswerth, W. S. Johnson, S. E. Schoenig, and G. C. van Kooten. "Costs and Losses Imposed on California Ranchers by Yellow Starthistle." *Rangeland Ecology & Management* 60 (2007): 369–77. doi:10.2111/1551-5028(2007)60[369:calioc]2.0.co;2.

30. K. F. Raffa, B. H. Aukema, B. J. Bentz, A. L. Carroll, J. A. Hicke, M. G. Turner, and W. H. Romme. "Cross-Scale Drivers of Natural Disturbances Prone to Anthropogenic Amplification: The Dynamics of Bark Beetle Eruptions." *BioScience* 58 (2008): 501–17. doi:10.1641/b580607.

31. Ibid.

32. Sue E. Moore and Henry P. Huntington. "Arctic Marine Mammals and Climate Change: Impacts and Resilience." *Ecological Applications* 18, no. sp2 (2008).

33. J. N. Mills, B. A. Ellis, K. T. McKee Jr., G. E. Calderon, J. I. Maiztegui, G. O. Nelson, . . . and J. E. Childs. "A Longitudinal Study of Junin Virus Activity in the Rodent Reservoir of Argentine Hemorrhagic Fever." *American Journal of Tropical Medicine and Hygiene* 47, no. 6 (1992): 749–63.

34. J. N. Mills and J. E. Childs. "Ecologic Studies of Rodent Reservoirs: Their Relevance for Human Health." *Emerging Infectious Diseases, Perspectives* 4, no. 4 (1998): 529–37.

35. Ibid.

CHAPTER 14

1. United Nations (UN). Universal Declaration of Human Rights, http://www.un.org/en/universal-declaration-human-rights/, December 10, 1948. Accessed October 5, 2016.

2. M. Robinson. "Foreword." *Health and Human Rights Journal* 16, no. 1 (2014): 4–7.

3. Resolution of the United Nations Human Rights Council: "Human Rights and Climate Change," http://www.ohchr.org/Documents/Issues/ClimateChange/A.HRC.RES.18.22.pdf?. Accessed March 16, 2017.

4. C. Williams. "Climate Change and the Right to Health." In *Global Climate Change and Human Health*, edited by G. Luber and J. Lemery, 601–15. San Francisco: John Wiley & Sons, 2015.

5. B. Cicin-Sain, R. W. Knecht, and N. Foster. "Trends and Future Challenges of U.S. National Ocean and Coastal Policy," Center for the Study of Marine Policy, University of Delaware, http://biotech.law.lsu.edu/cphl/noaa/ctrends-proceed-1999.pdf#page=28, January 22, 1999. Accessed October 5, 2016.

6. World Bank. "Poverty," http://www.worldbank.org/en/topic/poverty. Accessed November 28, 2016.

7. UN Development Program: Multidimensional Poverty Index, http://hdr.undp.org/en/content/multidimensional-poverty-index-mpi. Accessed November 28, 2016.

8. J. E. Cohen. "Human Population: The Next Half Century." *Science* 302, no. 5648 (2003): 1172–75.

9. K. Smith. "Climate Change, Violence, and the Afterlife. Health of People, Places and Planet," http://press-files.anu.edu.au/downloads/press/p320071/html/ch34.xhtml?referer=235&page=62, July 2015. Accessed October 5, 2016.

10. UN Committee on Economic, Social, and Cultural Rights. "General Comment 14," http://www.ohchr.org/Documents/Issues/Women/WRGS/Health/GC14.pdf, August 11, 2000. Accessed October 5, 2016.

11. "Malé Declaration on the Human Dimension of Global Climate Change," http://www.ciel.org/Publications/Male_Declaration_Nov07.pdf, November 14, 2007. Accessed October 5, 2016.

12. Williams, "Climate Change and the Right to Health."

CHAPTER 15

1. Centers for Disease Control and Prevention (CDC). "Trends in Current Cigarette Smoking among High School Students and Adults, United States, 1965–2014," http://www.cdc.gov/tobacco/data_statistics/tables/trends/cig_smoking/. Accessed September 30, 2016.

2. D. M. Burns, L. Lee, L. Z. Shen et al. "Cigarette Smoking Behavior in the United States," http://cancercontrol.cancer.gov/brp/tcrb/monographs/8/m8_2.pdf. Accessed September 30, 2016.

3. C. Bates and A. Rowell. "Tobacco Explained: The Truth about the Tobacco Industry . . . in Its Own Words," http://www.who.int/tobacco/media/en/Tobac coExplained.pdf. Accessed September 30, 2016.

4. CDC. "Smoking and Tobacco Use," http://www.cdc.gov/tobacco/data_statistics/fact_sheets/adult_data/cig_smoking/. Accessed October 12, 2016.

5. International Physicians for the Prevention of Nuclear War. "IPPNW: A Brief History," http://www.ippnw.org/history.html. Accessed October 5, 2016.

6. Jay Lemery. "Lessons From Dr. Strangelove." *Wilderness & Environmental Medicine* 23, no. 1 (2012): 2–4.

7. C. Hamilton. *Earthmasters: The Dawn of the Age of Climate Engineering*. New Haven, CT: Yale University Press, 2013.

8. Ibid.

9. N. Watts, N. W. Adger, P. Agnolucci et al. "Health and Climate Change: Policy Responses to Protect Public Health." *Lancet* 386, no. 10006 (2015): 1861–1914.

10. J. Lemery, C. Williams, and P. Farmer. "Editorial: The Great Procrastination." *Health & Human Rights: An International Journal* 16, no. 1 (2014): 1–3.

Bibliography

2014 National Climate Assessment, http://nca2014.globalchange.gov/highlights/report
-findings/extreme-weather. Accessed November 28, 2016.

Abdurahman M. El-Sayed, and Sandro Galea. "Climate Change and Population Mental Health." In *Global Climate Change and Human Health*, edited by G. Luber and J. Lemery. San Francisco: John Wiley & Sons, 2015, 317.

Ahern, M., S. Kovats, P. Wilkinson, R. Few, and F. Matthies. "Global Health Impacts of Floods: Epidemiologic Evidence." *Epidemiologic Reviews* 27 (January 2005): 36–46.

Alavanja, M. C., J. A. Hoppin, and F. Kamel. "Health Effects of Chronic Pesticide Exposure: Cancer and Neurotoxicity." *Annu. Rev. Public Health* 25 (2004): 155–97.

Allen, C. D., and D. D. Breshears. "Drought-Induced Shift of a Forest-Woodland Ecotone: Rapid Landscape Response to Climate Variation." *Proceedings of the National Academy of Sciences* 95 (1998): 14839–42. doi:10.1073/pnas.95.25.14839.

American College of Allergy, Asthma & Immunology. "The Year 2040: Double the Pollen, Double the Allergy Suffering," http://acaai.org/news/year-2040-double-pollen-double-allergy-suffering, 2014. Accessed September 26, 2016.

Anderson D. M., J. M. Burkholder, W. P. Cochlan et al. "Harmful Algal Blooms and Eutrophication: Examining Linkages from Selected Coastal Regions of the United States." *Harmful Algae* 8, no. 1 (2008): 39–53.

Annadurai K., R. Danasekaran, and G. Mani. "Global Eradication of Guinea Worm Disease: Toward a Newer Milestone." *J Res Med Sci* 19, no. 12 (2014): 1207–8.

Arctic Climate Impact Assessment (ACIA). *Impacts of a Warming Arctic: Arctic Climate Impact Assessment* (ACIA Overview Report). Cambridge: Cambridge University Press, 2004.

Aroian, K. J., and A. E. Norris. "Depression Trajectories in Relatively Recent Immigrants." *Comprehensive Psychiatry* 44, no. 5 (2003): 420–27.

Asher, M. Innes, et al. "Worldwide Time Trends in the Prevalence of Symptoms of Asthma, Allergic rhinoconjunctivitis, and Eczema in Childhood: ISAAC Phases One and Three Repeat Multicountry Cross-Sectional Surveys." *Lancet* 368, no. 9537 (2006): 733–43.

Assefa, F., M. Z. Jabarkhil, P. Salama, and P. Spiegel. "Malnutrition and Mortality in Kohistan District, Afghanistan, April 2001." *JAMA* 286, no. 21 (2001): 2723–28.

Auerbach, P. S. *Medicine for the Outdoors*. Philadelphia: Elsevier, 2016.

———. "Physicians and the Environment." *JAMA* 299, no. 8 (2008): 956–58.

Awasthi, M. D., A. K. Ahuja, and D. Sharma. "Contamination of Horticulture Ecosystem: Orchard Soil and Water Bodies with Pesticide Residues." Proceedings of National Symposium on Integrated Pest Management (IPM) in Horticultural Crops: New Molecules. Biopesticides and Environment, 2001.

Azevedo, S. M., W. W. Carmichael, E. M. Jochimsen, K. L. Rinehart, S. Lau, G. R. Shaw, and G. K. Eaglesham. "Human Intoxication by Microcystins during Renal Dialysis Treatment in Caruaru—Brazil." *Toxicology* 181 (2002): 441–46.

Backer, L. C. "Effects of Climate Change on Noninfectious Waterborne Threats." In *Global Climate Change and Human Health*, edited by G. Luber and J. Lemery, 171–93. San Francisco: John Wiley & Sons, 2015.

Baden, D., L. Fleming, and J. Bean. "Marine Toxins." In *Handbook of Clinical Neurology: Intoxications of the Nervous System Part II. Natural Toxins and Drugs*, edited by F. A. deWolff, 141–75. Amsterdam: Elsevier Press, 1995.

Baillie, J., C. Hilton-Taylor, and S. N. Stuart. "2004 IUCN Red List of Threatened Species: A Global Species Assessment." International Union for Conservation of Nature and Natural Resources, 2004.

Baily, S. W. "Climate Change and Decreasing Herbicide Persistence." *Pest Management Science* 60 (2004): 158–62.

Barford, E. "Crop Pests Advancing with Global Warming." *Nature* (September 2013). doi:10.1038/nature.2013.13644.

Barnett, J., and M. Webber. "Accommodating Migration to Promote Adaption to Climate Change." Policy Research Working Paper 5270, April 2010. http://poseidon01.ssrn.com/delivery.php ?ID=05808407002000911210307612710210201102801502600206002309601900805806104204504700301106909311809609407901706103210708610612209812002009607008712311801712109601609706512712509612707511909402&EXT=pdf. Accessed October 1, 2016.

Barnett, T. P., D. W. Pierce, H. G. Hidalgo, C. Bonfils, B. D. Santer, T. Das, G. Bala, A. W. Wood, T. Nozawa, A. A. Mirin, D. R. Cayan, and M. D. Dettinger. "Human-Induced Changes in the Hydrology of the Western United States." *Science* 319 (2008): 1080–83. doi:10.1126/science.1152538.

Bates, C., and A. Rowell. "Tobacco Explained; The Truth about the Tobacco Industry . . . in Its Own Words," http://www.who.int/tobacco/media/en/TobaccoExplained.pdf. Accessed September 30, 2016.

Bates, D. V., and R. Sizto. "Air Pollution and Hospital Admissions in Southern Ontario: The Acid Summer Haze Effect." *Environmental Research* 43, no. 2 (1987): 317–31.

Beard, C. D., J. Garofalo, and K. Gage. "Climate and Its Impacts on Vector-Borne and Zoonotic Disease." In *Global Climate Change and Human Health*, edited by G. Luber and J. Lemery, 221–66. San Francisco: John Wiley & Sons, 2015.

Beck, P. S. A., G. P. Juday, C. Alix, V. A. Barber, S. E. Winslow, E. E. Sousa, P. Heiser, J. D. Herriges, and S. J. Goetz. "Changes in Forest Productivity across Alaska Consistent with Biome Shift." *Ecology Letters* 14 (2011): 373–79. doi:10.1111/j.1461-0248.2011.01598.x.

Beckage, B., B. Osborne, D. G. Gavin, C. Pucko, T. Siccama, and T. Perkins. "A Rapid Upward Shift of a Forest Ecotone during 40 Years of Warming in the Green Mountains of Vermont." *Proceedings of the National Academy of Sciences* 105 (2008): 4197–4202. doi:10.1073/pnas.0708921105.

Bell, M. L., S. L. McDermott, J. M. Samet, J. M. Zeger, and F. Dominici. "Ozone and Mortality in 95 U.S. Urban Communities, 1987 to 2000." *JAMA* 292 (2004): 2372–78.

Bellard, C., C. Bertelsmeier, P. Leadley, W. Thuiller, and F. Courchamp. "Impacts of Climate Change on the Future of Biodiversity." *Ecology Letters* 15, no. 4 (2012): 365–77.

Beniston, M. "The 2003 Heat Wave in Europe: A Shape of Things to Come? An Analysis Based on Swiss Climatological Data and Model Simulations." *Geophysical Research Letters* 31, no. 2 (2004).

Bennett, Bradley C. "Doctrine of Signatures: An Explanation of Medicinal Plant Discovery or Dissemination of Knowledge?" *Economic Botany* 61, no. 3 (2007): 246–55.

Bennish, M. L., and B. J. Wojtyniak. "Mortality Due to Shigellosis: Community and Hospital Data. *Rev Infect Dis* 13, no. 4 (March–April 1991): S245–51.

Berman, J. D., N. Fann, J. W. Hollingsworth, K. E. Pinkerton, W. N. Rom, A. M. Szema, P. N. Breysse, R. H. White, and F. C. Curriero. "Health Benefits from Large-Scale Ozone Reduction in the United States." *Environmental Health Perspectives* 120 (2012): 1404–10. http://dx.doi.org/10.1289/ehp.1104851.

Blake, E. S., T. B. Kimberlain, R. J. Berg, J. P. Cangialosi, and J. L. Beven II. "Tropical Cyclone Report: Hurricane Sandy." *National Hurricane Center* 12 (2013): 1–10.

Booth, S., and D. Zeller. "Mercury, Food Webs, and Marine Mammals: Implications of Diet and Climate Change for Human Health." *Environmental Health Perspectives* 113 (2005): 521–26.

Bradley, M., R. Shakespeare, A. Ruwende, M. E. Woolhouse, E. Mason, and A. Munatsi. "Epidemiological Features of Epidemic Cholera (El Tor) in Zimbabwe." *Transactions of the Royal Society of Tropical Medicine & Hygiene* 90, no. 4 (1996): 378–82.

Brownstein, J. S., T. R. Holford, and D. Fish. "Effect of Climate Change on Lyme Disease Risk in North America." *Ecohealth* 2, no. 1 (2005): 38–46.

Burkle, Frederick. "Foundations for Global Health Responders." Massive Open Online Course. (Coursera, 2015), https://www.coursera.org/learn/ghresponder.

Burns, D. M., L. Lee, L. Z. Shen et al. "Cigarette Smoking Behavior in the United States," http://cancercontrol.cancer.gov/brp/tcrb/monographs/8/m8_2.pdf. Accessed September 30, 2016.

Callaway, E. "Pathogen Genome Tracks Irish Potato Famine Back to Its Roots." *Nature* (May 2013). doi:10.1038/nature.2013.13021.

Caller, T. A., J. W. Doolin, J. F. Haney et al. "A Cluster of Amyotrophic Lateral Sclerosis in New Hampshire: A Possible Role for Toxic Cyanobacteria Blooms." *Amyotrophic Lateral Sclerosis* 10, no. 2 (2009): 101–8.

Cardozo, B. L., O. O. Bilukha, C. A. Gotway Crawford et al. "Mental Health, Social Functioning, and Disability in Postwar Afghanistan." *JAMA* 292 (2004): 575–84.

Carling, P. C., L. A. Bruno-Murtha, and J. K. Griffiths. "Cruise Ship Environmental Hygience and the Risk of Norovirus Infection Outbreaks: An Objective Assessment of 56 Vessels over 3 Years." *Clinical Infectious Diseases* 49, no. 9 (2009): 1312–17.

Celenza, A., J. Fothergill, E. Kupek, and R. J. Shaw. "Thunderstorm Associated Asthma: A Detailed Analysis of Environmental Factors." *BMJ* 312, no. 7031 (1996): 604–7.

Center for Earth and Environmental Science. "What Causes Algal Bloom?" http://www.cees.iupui.edu/research/algaltoxicology/bloomfactors. Accessed October 12, 2016.

Centers for Disease Control and Prevention (CDC). "Mortality during a Famine—Gode District, Ethiopia, July 2000." *Morbidity and Mortality Weekly Report* 50, no. 15 (2001): 285–88.

———. "Increase in coccidioidomycosis—Arizona, 1998–2001." *Morbidity and Mortality Weekly Report* 52, no. 6 (2003): 109–12.

———. "Climate Effects on Health," http://www.cdc.gov/climateandhealth/effects/, December 22, 2014. Accessed June 13, 2016.

———. "Coccidioidomycosis," http://www.cdc.gov/fungal/diseases/coccidioidomycosis/. Accessed November 28, 2016.

———. "Dengue," https://www.cdc.gov/dengue/. Accessed October 12, 2016.

———. "Dracunculiasis," http://www.cdc.gov/parasites/guineaworm/index.html. Accessed October 12, 2016.

———. "Haff Disease Associated with Eating Buffalo Fish—United States, 1997." *Morbidity and Mortality Weekly Report* 47(50) (1998): 1091–93.

———. "Lyme Disease," http://www.cdc.gov/lyme/. Accessed October 12, 2016.

———. http://www.cdc.gov/aging/pdf/disaster_planning_goal.pdf. Accessed October 12, 2016.

———. "Norovirus Outbreak among Evacuees from Hurricane Katrina—Houston, Texas, September 2005," https://www.cdc.gov/mmwr/preview/mmwrhtml/mm5440a3.htm, October 14, 2005. Accessed October 5, 2016.

———. "Pfiesteria," http://www.cdc.gov/hab/pfiesteria/pdfs/about.pdf. Accessed October 12, 2016.

———. "Shigella," https://www.cdc.gov/shigella/. Accessed October 12, 2016.

———. "Smoking and Tobacco Use," http://www.cdc.gov/tobacco/data_statistics/fact_sheets/adult_data/cig_smoking/. Accessed October 12, 2016.

———. "Typhoid Fever," http://www.cdc.gov/typhoid-fever/. Accessed October 12, 2016.

———. "Trends in Current Cigarette Smoking among High School Students and Adults, United States, 1965–2014," http://www.cdc.gov/tobacco/data_statistics/tables/trends/cig_smoking/, March 30, 2016. Accessed September 30, 2016.

———. "West Nile Virus," https://www.cdc.gov/westnile/. Accessed October 12, 2016.

———. "Zika Virus," https://www.cdc.gov/zika/healtheffects/index.html. Accessed October 12, 2016.

Centers for Disease Control and Prevention (CDC) Climate & Health Program, https://www.cdc.gov/climateandhealth/.

Centers for Research on the Epidemiology of Disasters. *EM-DAT: The International Disaster Database*, http://www.emdat.be/disaster-trends. Brussels: Ecole de santé publique, Université catholique de Louvain, 2009.

Chase, J., and T. Knight. "Drought-Induced Mosquito Outbreaks in Wetlands." *Ecology Letters* 6 (2003): 1017–24.

Chen, M., C. Lin, and Y. Wu et al. "Effects of Extreme Precipitation to the Distribution of Infectious Diseases in Taiwan, 1994–2008." *PLoS One* 7, no. 6 (2012): e34651.

Chitsulo, L., D. Engels, A. Montresor, and L. Savioli. "The Global Status of Schistosomiasis and Its Control." *Acta Trop* 77 (2000): 41–51. doi:10.1016/S0001-706X(00)00122-4.

Chivian, E., and A. Bernstein. "How Our Health Depends on Biodiversity," http://www.chgeharvard.org/sites/default/files/resources/182945%20HMS%20Biodiversity%20booklet.pdf. Center for Health and the Global Environment, Harvard Medical School. Accessed October 8, 2016.

———. *Sustaining Life*. Oxford: Oxford University Press, 2008, 117.

Chmid, D., I. Lederer, P. Much et al. "Outbreak of Norovirus Infection Associated with Contaminated Flood Water, Salzburg 2005." *Eurosurveillance* 10, no. 24 (2005): 2727.

Choi, M., F. C. Curriero, M. Johantgen et al. "Association between Ozone and Emergency Department Visits: An Ecological Study." *International Journal of Environmental Health Research* 21 (2011): 201–21. doi:10.1080/096-3123.2010.533366.

Christidis, N., G. S. Jones, and P. A. Stott. "Dramatically Increasing Chance of Extremely Hot Summers since the 2003 European Heatwave." *Nature Climate Change* 5, no. 1 (2015): 46–50.

Church, J. A., P. U. Clark, A. Cazenave et al. "Intergovernmental Panel on Climate Change." Fifth Assessment Report (AR5), chap. 13, Sea Level Change, https://www.ipcc.ch/pdf/assessment-report/ar5/wg1/WG1AR5_Chapter13_FINAL.pdf. Accessed October 1, 2016.

Chusid, M. J., D. C. Dale, B. C. West, and S. M. Wolff. "The Hypereosinophilic Syndrome: Analysis of Fourteen Cases with Review of the Literature." *Medicine* 54, no. 1 (1975): 1–27.

CIA World Factbook: "Tuvalu," https://www.cia.gov/library/publications/the-world-factbook/geos/tv.html. Accessed March 16, 2017.

Cicin-Sain, B., R. W. Knecht, and N. Foster. "Trends and Future Challenges of U.S. National Ocean and Coastal Policy." Center for the Study of Marine Policy, University of Delaware. National Ocean Service, NOAA, January 22, 1999, http://biotech.law.lsu.edu/cphl/noaa/ctrends-proceed-1999.pdf#page=28. Accessed October 5, 2016.

Clark, D., and A. Garner. "The First 1000 Days: Do YOU See What We See?" *American Academy of Pediatrics*, https://www.aap.org/en-us/advocacy-and-policy/aap-health-initiatives/EBCD/Documents/ALF-1000days.pdf.

Cline, W. R. *Global Warming and Agriculture: Impact Estimates by Country*. Washington, DC: Center for Global Development and Peterson Institute for International Economics, 2007.

Cohen, J. E. "Human Population: The Next Half Century." *Science* 302, no. 5648 (2003): 1172–75.

Collie, J. S., A. D. Wood, and H. P. Jeffries. "Long-Term Shifts in the Species Composition of a Coastal Fish Community." *Canadian Journal of Fisheries and Aquatic Sciences* 65 (2008): 1352–65. doi:10.1139/F08-048.

Cook, B., R. Miller, and R. Seager. "Did Dust Storms Make the Dust Bowl Drought Worse?" http://www.ldeo.columbia.edu/res/div/ocp/drought/dust_storms.shtml. Lamont-Doherty Earth Observatory, The Earth Institute at Columbia University, 2007.

Cook, B. I., K. J. Anchukaitis, R. Touchan, D. M. Meko, and E. R. Cook. "Spatio-temporal Drought Variability in the Mediterranean over the Last 900 Years." *Journal of Geophysical Research: Atmospheres* (2016).

Cook, B. I., T. R. Ault, and J. E. Smerdon. "Unprecedented 21st Century Drought Risk in the American Southwest and Central Plains." *Science Advances* 1, no. 1 (2015): e1400082.

Cox, P. A., R. Richer, J. S. Metcalf, S. A. Banack, G. A. Codd, and W. G. Bradley. "Cyanobacteria and BMAA Exposure from Desert Dust: A Possible Link to Sporadic ALS among Gulf War Veterans." *Amyotrophic Lateral Sclerosis* 10, supp 2 (2009): 109–17.

Cragg, G. M., and D. J. Newman. "Natural Products: A Continuing Source of Novel Drug Leads." *Biochimica et Biophysica Acta (BBA)—General Subjects* 1830, no. 6 (2013): 3670–95.

Crump, J. A., S. P. Luby, and E. D. Mintz. "The Global Burden of Typhoid Fever." *Bull World Health Organ* 82 (2004): 346–53.

Cryptosporidiosis Surveillance—United States, 2011–2012. Centers for Disease Control and Prevention, *Morbidity and Mortality Weekly Report*, http://www.cdc.gov/mmwr/preview/mmwrhtml/ss6403a1.htm, May 1, 2015. Accessed May 6, 2016.

Damalas, C. A., and I. G. Eleftherohorinos. "Pesticide Exposure, Safety Issues, and Risk Assessment Indicators." *Int J Environ Res Public Health* 8, no. 5 (May 2011): 1402–19. doi:10.3390/ijerph8051402.

D'Amato, G., G. Liccardi, M. D'Amato, and M. Cazzola. "Outdoor Air Pollution, Climatic Changes and Allergic Bronchial Asthma." *European Respiratory Journal* 20, no. 3 (2002): 763–76.

Dart, R. C. *Medical Toxicology*, 3rd ed. Philadelphia: Lippincott Williams & Wilkins, 2004.

Davies, R. H., and C. Wra. "Seasonal Variations in the Isolation of Salmonella typhimurium, Salmonella enteritidis, Bacillus cereus and Clostridium perfringens from Environmental Samples." *Journal of Veterinary Medicine*, ser. B, 43, nos. 1–10 (1996): 119–127.

de Silva, N. R., S. Brooker, P. J. Hotez, A. Montresor, D. Engles, and L. Savioli. "Soil-Transmitted Helminth Infections: Updating the Global Picture." *Trends in Parasitology* 19 (2003): 547–51.

Dennekamp, M., and M. J. Abramson. "The Effects of Bushfire Smoke on Respiratory Health." *Respirology* 16 (2011): 198–209.

DesGranges, J. L., J. Rodrigue, B. Tardif, and M. Laperle. "Mercury Accumulation and Biomagnification in Ospreys (Pandion haliaetus) in the James Bay and Hudson Bay Regions of Quebec." *Archives of Environmental Contamination and Toxicology* 35, no. 2 (1998): 330–41.

Dhainaut, J. F., Y. E. Claessens, C. Ginsburg, and B. Riou. "Unprecedented Heat-Related Deaths during the 2003 Heat Wave in Paris: Consequences on Emergency Departments." *Critical Care* 8, no. 1 (2003): 1.

Dillingham, R. A., A. A. Lima, and R. L. Guerrant. "Cryptosporidiosis: Epidemiology and Impact." *Microbes and Infection* 4, no. 10 (2002): 1059–66.

Dorfman, M., N. Stoner, and M. Merkel. "Swimming in Sewage." Natural Resources Defense Council, https://www.nrdc.org/sites/default/files/sewage.pdf, February 2004. Accessed June 9, 2016.

Doyle, A., D. Barataud, A. Gallay et al. "Norovirus Foodborne Outbreaks Associated with the Consumption of Oysters from the Etang de Thau, France, December 2012." *Euro Surveill* 9, no. 3 (March 2004): 24–26.

"Drought in Eastern Mediterranean Worst of Past 900 Years." NASA Global Climate Change, http://climate.nasa.gov/news/2408/, March 1, 2016. Accessed May 6, 2016.

Dukes, J. S., N. R. Chiariello, S. R. Loarie, and C. B. Field. "Strong Response of an Invasive Plant Species (Centaurea solstitialis L.) to Global Environmental Changes." *Ecological Applications* 21 (2011): 1887–94. doi:10.1890/11-0111.1.

Eagle, A. J., M. E. Eiswerth, W. S. Johnson, S. E. Schoenig, and G. C. van Kooten. "Costs and Losses Imposed on California Ranchers by Yellow Starthistle." *Rangeland Ecology & Management* 60 (2007): 369–77. doi:10.2111/1551-5028(2007)60[369:ca lioc]2.0.co;2.

Easterling, W. E., P. Aggarwal, P. Batima, K. Brander, L. Erda, S. Howeden, A. Kirilenko et al. "Food, Fibre and Forest Products. Climate Change 2007: Impacts, Adaptation and Vulnerability," Contribution of Working Group II to the Fourth Assessment Report of the Intergovernmental Panel on Climate Change, edited by M. L. Parry, 273–313. Cambridge: Cambridge University Press, 2007.

Ecosystem Services, Biodiversity and Human Health, Harvard School of Public Health website, http://www.chgeharvard.org/topic/ecosystem-services, 2012–2016. Accessed June 4, 2016.

Edwards, B., M. Gray, and B. Hunter. "Social and Economic Impacts of Drought on Farm Families and Rural Communities: Submission to the Productivity Commission's Inquiry into Government Drought Support." Melbourne: Australian Institute of Family Studies, 2008.

Edwards, M., D. G. Johns, S. C. Leterme, E. Svendsen, and A. J. Richardson. "Regional Climate Change and Harmful Algal Blooms in the Northeast Atlantic." *Limnology and Oceanography* 51 (2006): 820–29.

Efferth, T., A. Benakis, M. R. Romero, M. Tomicic, R. Rauh, D. Steinbach, . . . and M. Marschall. "Enhancement of Cytotoxicity of Artemisinins toward Cancer Cells by Ferrous Iron." *Free Radical Biology and Medicine* 37, no. 7 (2004): 998–1009.

Effler, E., M. Isaacson, L. Arntzen, R. Heenan, P. Canter, T. Barrett, L. Lee, C. Mambo, W. Levine, A. Zaidi, and P. M. Griffin. "Factors Contributing to the

Emergence of Escherichia coli O157 in Africa." *Emerging Infectious Diseases* 7, no. 5 (2001): 812–19.

El-Sayed, A. M., and S. Galea. "Climate Change and Population Mental Health." In *Global Climate Change and Human Health*, edited by G. Luber and J. Lemery, 311–31. San Francisco: John Wiley & Sons, 2015.

Environmental Protection Agency (EPA), "Health Risks of Human Exposure to Wastewater." https://nepis.epa.gov/Exe/ZyNET.exe/20016SRO.TXT?ZyAction D=ZyDocument&Client=EPA&Index=1981+Thru+1985&Docs=&Query=&Ti me=&EndTime=&SearchMethod=1&TocRestrict=n&Toc=&TocEntry=&QFiel d=&QFieldYear=&QFieldMonth=&QFieldDay=&IntQFieldOp=0&ExtQFieldO p=0&XmlQuery=&File=D%3A%5Czyfiles%5CIndex%20Data%5C81thru85%5C Txt%5C00000013%5C20016SRO.txt&User=ANONYMOUS&Password=anony mous&SortMethod=h%7C-&MaximumDocuments=1&FuzzyDegree=0&ImageQ uality=r75g8/r75g8/x150y150g16/i425&Display=hpfr&DefSeekPage=x&SearchB ack=ZyActionL&Back=ZyActionS&BackDesc=Results%20page&MaximumPages =1&ZyEntry=1&SeekPage=x&ZyPURL. Accessed November 28, 2016.

———. "Murphy Oil Spill Factsheet," February 2006, http://www.columbia.edu/ itc/journalism/cases/katrina/Federal%20Government/Environmental%20Protec- tion%20Agency/Murphy%20Oil%20Spill%20Fact%20Sheet%20Feb%202006.pdf. Accessed October 12, 2016.

———. https://www.epa.gov/mercury/how-people-are-exposed-mercury. Accessed November 28, 2016.

Epstein, Paul R., and Evan Mills. "Climate Change Futures: Health, Ecological and Economic Dimensions." The Center for Health and the Global Environment, Har- vard Medical School, 2005.

Ernesto, M., A. P. Cardoso, D. Nicala, E. Mirione, F. Massaza, J. Cliff, . . . and J. H. Bradbury. "Persistent Konzo and Cyanogen Toxicity from Cassava in Northern Mozambique." *Acta Tropica* 82, no. 3 (2002): 357–62.

Esch, K. J., and C. A. Petersen. "Transmission and Epidemiology of Zoonotic Proto- zoal Diseases of Companion Animals." *Clin Microbiol Rev* 26, no. 1 (January 2013): 58–85.

Evans, G. W. "The Built Environment and Mental Health." *Journal of Urban Health* 80, no. 4 (2003): 536–55.

Falconer, I. "Algal Toxins and Human Health." In *The Handbook of Environmental Chemistry*, vol. 5, edited by J. Hrub, 53–82. Berlin: Springer-Verlag, 1998.

———. "Measurement of Toxins from Blue-Green Algae in Water and Foodstuffs." In *Algal Toxins in Seafood and Drinking Water*, edited by I. R. Falconer, 165–75. Orlando, FL: Academic Press, 1993.

Fann, Neal, et al. "Estimating the National Public Health Burden Associated with Exposure to Ambient PM2. 5 and Ozone." *Risk Analysis* 32, no. 1 (2012): 81–95.

Fernando, M. A. "Effect of Ascaris lumbricoides Infestation on Growth of Children." *Indian Pediatrics* 20, no. 10 (1983): 721.

Field, L., J. W. Kern, and L. B. Rosman. "Re-visiting Projections of PCBs in the Lower Hudson River Fish Using Model Emulation." *Science of the Total Environment* (July 2016): 557–58, 489–501. doi:10.1016/j.scitotenv.2016.02.072.

Fiore, A. M., V. Naik, D. V. Spracklen, A. Steiner, N. Unger, M. Prather, . . . and S. Dalsøren. "Global Air Quality and Climate." *Chemical Society Reviews* 41, no. 19 (2012): 6663–83.

"First 1,000 Days Last Forever: Scaling Up Nutrition for a Just World," UNICEF, http://www.unicef.org/tajikistan/Op-Ed_in_support_of_UNICEF_global_nutrition_report_adopted_TJK_facts_ENG.pdf, 2012. Accessed June 4, 2016.

Fitzpatrick, K., and M. LaGory. *Unhealthy Places: The Ecology of Risk in the Urban Landscape*. New York: Routledge, 2002.

Fleming, L. E., D. Kirkpatrick, L. Backer, J. Bean, A. Wanner, D. Dalpra, R. Tamer et al. "Initial Evaluation of the Effects of Aerosolized Florida Red Tide Toxins (Brevetoxins) in Persons with Asthma." *Environmental Health Perspectives* 113 (2005): 650–57.

"Floods in the WHO European Region: Health Effects and Their Preventions." World Health Organization, https://www.researchgate.net/publication/275345314_Floods_and_Public_Health_Consequences_Prevention_and_Control_Measures, January 9, 2016. Accessed May 7, 2016.

Fouillet, A., G. Rey, V. Wagner, K. Laaidi, P. Empereur-Bissonnet, A. Le Tertre, . . . and E. Jougla. "Has the Impact of Heat Waves on Mortality Changed in France since the European Heat Wave of Summer 2003? A Study of the 2006 Heat Wave." *International Journal of Epidemiology* 37, no. 2 (2008): 309–17.

Freye, H. B., J. King, and C. M. Litwin. "Variations of Pollen and Mold Concentrations in 1998 during the Strong El Nino Event of 1997–1998 and Their Impact on Clinical Exacerbations of Allergic Rhinitis, Asthma, and Sinusitis." *Allergy and Asthma Proceedings* 22 (2001): 239–47.

Friedman, M. S., K. E. Powell, L. Hutwagner, L. M. Graham, and W. G. Teague. "Impact of Changes in Transportation and Commuting Behaviors during the 1996 Summer Olympic Games in Atlanta on Air Quality and Childhood Asthma." *JAMA* 285, no. 7 (2001): 897–905.

Fukiki, H., M. Suganuma, H. Suguri et al. "Diarrhetic Shellfish Toxin, Dinophysistoxin-1, Is a Potent Tumor Promoter on Mouse Skin." *Cancer Science* 79, no. 10 (1988): 1089–93.

Fuller, R. B., and J. Snyder. *Operating Manual for Spaceship Earth*. Carbondale: Southern Illinois University Press, 1969.

Gagnon, A. S., K. E. Smoyer-Tomic, and A. B. Bush. "The El Nino Southern Oscillation and Malaria Epidemics in South America." *International Journal of Biometeorology* 46, no. 2 (2002): 81–89.

Ganesh, S., and R. J. Cruz Jr. "Strongyloidiasis; A Multifaceted Disease." *Gastroenterol Hepatol* 7, no. 3 (2011): 194–96.

Getahun, H., A. Mekonnen, R. TekleHaimanot, and F. Lambein. "Epidemic of Neurolathyrism in Ethiopia." *Lancet* 354, no. 9175 (1999): 306–7.

Gevao, B., K. T. Semple, and K. C. Jones. "Bound Pesticide Residues in Soils: A Review." *Environmental Pollution* 108, no. 1 (2000): 3–14.

"Giardiasis Surveillance—United States, 2011–2012." MMWR Centers for Disease Control and Prevention, http://www.cdc.gov/mmwr/preview/mmwrhtml/ss6403a2.htm, May 1, 2015. Accessed May 6, 2016.

Githeko, A. K., and W. Ndegwa. "Predicting Malaria Epidemics in the Kenyan High-lands Using Climate Data: A Tool for Decision Makers." *Global Change and Human Health* 2, no. 1 (2001): 54–63.

Gleick, P. H. "The Syrian Conflict and the Role of Water." In *The World's Water*, edited by P. H. Gleick, 147–51. Washington, DC: Island Press/Center for Resource Economics, 2014.

"Global Epidemics and Impact of Cholera." World Health Organization, http://www.who.int/topics/cholera/impact/en, 2016. Accessed May 6, 2016.

Gomez, S. R., R. A. Parker, J. A. Dosman, and H. H. McDuffie. "Respiratory Health Effects of Alkali Dust in Residents Near Desiccated Old Wives Lake." *Archives of Environmental Health* 47, no. 5 (1992): 364–69.

Grant, K. S., T. M. Burbachker, E. M. Faustman, and L. Grattan. "Domoic Acid: Neurobehavioral Consequences of Exposure to a Prevalent Marine Biotoxin." *Neurotoxicol Teratol* 32, no. 2 (2010): 132–41.

Griffin, D., and K. J. Anchukaitis. "How Unusual Is the 2012–2014 California Drought?" *Geophysical Research Letters*, American Geophysical Union, 2014. doi:10.1002/2014GL062433.

Griggs, D. "Climate Policy: Streamline IPCC Reports." *Nature* 508, no. 7495 (2014): 171–73.

Grove, N. J., and A. B. Zwi. "Our Health and Theirs: Forced Migration, Othering, and Public Health." *Social Science & Medicine* 62, no. 8 (2006): 1931–42.

Grulk, N., and S. Schilling. "Air Pollution and Climate Change," http://www.fs.fed.us/ccrc/topics/air-pollution.shtml. Washington, DC: U.S. Department of Agriculture, Forest Service, Climate Change Resource Center, 2008.

Guan, W. J., X. Y. Zheng, K. F. Chung, and N. S. Zhong. "Impact of Air Pollution on the Burden of Chronic Respiratory Diseases in China: Time for Urgent Action." *Lancet* 388, no. 10054 (2016): 1939–51.

Gubler, Duane J. "TMH Dengue, Urbanization and Globalization: The Unholy Trinity of the 21st Century." *Tropical Medicine and Health* 39, no. 4 Supplement (2011): 3–11.

Hales, S., P. Weinstein, and A. Woodward. "Ciguatera (Fish Poisoning), El Niño, and Pacific Sea Surface Temperatures." *Ecosystem Health* 5, no. 1 (1999): 20–25.

Hamilton, C. *Earthmasters: The Dawn of the Age of Climate Engineering*. New Haven, CT: Yale University Press, 2013.

Harlan, S. L., A. J. Brazel, L. Prashad, W. L. Stefanov, and L. Larsen. "Neighborhood Microclimates and Vulnerability to Heat Stress." *Social Science & Medicine* 63, no. 11 (2006): 2847–63.

Hay, S. I., J. Cox, D. J. Rogers, S. E. Randolph, D. I. Stern, G. D. Shanks, . . . and R. W. Snow, "Climate Change and the Resurgence of Malaria in the East African Highlands." *Nature* 415, no. 6874 (2002): 905–9.

"Healthy Homes Issues: Pesticides in the Home—Use, Hazards, and Integrated Pest Management," http://portal.hud.gov/hudportal/documents/huddoc?id=DOC_12484.pdf, March 2006. Accessed June 4, 2016.

Hlavsa, M. C., V. A. Roberst, A. M. Kahler et al. "Outbreaks of Illness Associated with Recreational Water—United States, 2011–2012." *Morbidity and Mortality Weekly Report*, Centers for Disease Control and Prevention, http://www.cdc.gov/mmwr/preview/mmwrhtml/mm6424a4.htm, June 26, 2015. Accessed May 6, 2016.

Hoegh-Guldberg, O., P. J. Mumby, A. J. Hooten, R. S. Steneck, P. Greenfield, E. Gomez, . . . and N. Knowlton. "Coral Reefs under Rapid Climate Change and Ocean Acidification." *Science* 318, no. 5857 (2007): 1737–42.

Hsiang, S. M., M. Burke, and E. Miguel. "Quantifying the Influence of Climate on Human Conflict." *Science* 341, no. 6151 (2013): 1235367.

Hsu, E. "Reflections on the 'Discovery' of the Antimalarial Qinghao." *British Journal of Clinical Pharmacology* 61, no. 6 (2006): 666–70.

Idso, S. B., and K. E. Idso. "Effects of Atmospheric Carbon Dioxide Enrichment on Plant Constituents Related to Animal and Human Health." *Environmental and Experimental Botany* 45 (2001): 179–99.

Imhoff, M. L., L. Bounoua, T. Ricketts, C. Loucks, R. Harriss, and W. T. Lawrence. "Global Patterns in Human Consumption of Net Primary Production." *Nature* 42, no. 6994 (2004): 870–73.

"Impacts of Ocean Acidification on the Reef," Australian Government Great Barrier Reef Marine Park Authority, http://www.gbrmpa.gov.au/managing-the-reef/threats-to-the-reef/climate-change/how-climate-change-can-affect-the-reef/ocean-acidification, 2016. Accessed June 4, 2016.

Intergovernmental Panel on Climate Change. Assessment Report 5, chap. 9: Human Health. Working Group II: Impacts, Adaptation and Vulnerability. 9.7.2.1 Modeling the Impact of Climate Change on Dengue, http://www.ipcc.ch/ipccreports/tar/wg2/index.php?idp=361, 2001. Accessed October 8, 2016.

———. "Climate Change and Water," http://ipcc.ch/pdf/technical-papers/climate-change-water-en.pdf. Accessed December 27, 2016.

———. Fifth Assessment Report (AR5), https://www.ipcc.ch/report/ar5/. Accessed October 5, 2016.

International Physicians for the Prevention of Nuclear War. "IPPNW: A Brief History," http://www.ippnw.org/history.html. Accessed October 5, 2016.

International Year of Sanitation 2008: Overview. UNICEF, http://esa.un.org/iys/docs/IYS%20PRESS%20KIT.pdf, 2008. Accessed May 7, 2016.

IPPC Technical Paper VI. "Climate Change and Water," http://ipcc.ch/pdf/technical-papers/climate-change-water-en.pdf. Accessed April 10, 2016.

Isakbaeva, E. T., M. A. Widdowson, R. S. Beard, S. N. Bulens, J. Mullins, S. S. Monroe, . . . and R. I. Glass. "Norovirus Transmission on Cruise Ship." *Emerg Infect Dis* 11, no. 1 (2005): 154–58.

Ives, M., and J. Haner. "A Remote Pacific Nation, Threatened by Rising Seas." *New York Times*, July 2, 2016, http://www.nytimes.com/2016/07/03/world/asia/climate-change-kiribati.html?_r=0.

Jacob, D. J., and D. A. Winner. "Effect of Climate Change on Air Quality." *Atmospheric Environment* 43, no. 1 (2009): 51–63.

Jeffery, B., T. Barlow, K. Moizer, S. Paul, and C. Boyle. "Amnesic Shellfish Poison." *Food and Chemical Toxicology* 42, no. 4 (2004): 545–57.

Jerrett, M., R. T. Burnett, C. A. Pope III, K. Ito, G. Thurston, D. Krewski, Y. Shi et al. "Long-Term Ozone Exposure and Mortality." *New England Journal of Medicine* 360, no. 11 (2009): 1085–95. doi:10.1056/NEJMoa0803894.

Jeyaratnam, J. "Health Problems of Pesticide Usage in the Third World." *British Journal of Industrial Medicine* 42 (1985a): 505–6.

Johnson, B. J., and M. V. K. Sukhdeo. "Drought-Induced Amplification of Local and Regional West Nile Virus Infection Rates in New Jersey." *Journal of Medical Entomology* 50, no. 1 (2013): 195–204.

Karaolis, D. R., R. Lan, and P. R. Reeves. "The Sixth and Seventh Cholera Pandemics Are Due to Independent Clones Separately Derived from Environmental, Nontoxigenic, non-O1 Vibrio Cholera." *J Bacteriolm* 177, no. 11 (June 1995): 3191–98.

Kassalik, M., and K. Mönkemüller. "Strongyloides stercoralis hyperinfection Syndrome and Disseminated Disease." *Gastroenterol Hepatol* 7 (2011): 766–68.

Keim, M. E. "Extreme Weather Events; The Role of Public Health in Disaster Risk Reduction as a Means for Climate Change Adaption." In *Global Climate Change and Human Health*, edited by G. Luber and J. Lemery, 35–76. San Francisco: John Wiley & Sons, 2015.

Kelly, A. E., and M. L. Goulden. "Rapid Shifts in Plant Distribution with Recent Climate Change." *Proceedings of the National Academy of Sciences* 105 (2008): 11823–26. doi:10.1073/pnas.0802891105.

Kernéis, S., P. J. Guerin, L. von Seidlein et al. "A Look Back at an Ongoing Problem: Shigella dysenteriae Type 1 Epidemics in Refugee Settings in Central Africa (1993–1995)." Edited by D. J. Diemert. *PLoS ONE* 4, no. 2 (2009): e4494. doi:10.1371/journal.pone.0004494.

Kessler, R. C., S. Aguilar-Gaxiola, J. Alonso et al. "The Global Burden of Mental Disorders: An Update from the WHO World Mental Health (WMH) Surveys." *Epidemiol Psichiatr Soc* 18, no. 1 (2009): 23–33.

Kessler, R. C., S. Galea, M. J. Gruber et al. "Trends in Mental Illness and Suicidality after Hurricane Katrina." *Mol Psychiatry* 13, no. 4 (2008): 374–84.

Kessler, R. C., S. Galea, R. T. Jones et al. "Mental Illness and Suicidality after Hurricane Katrina." *Bulletin of the World Health Organization*, http://www.who.int/bulletin/volumes/84/10/06-033019.pdf. doi:10.2471/BLT.06.033019.

Kheir, Musa M., et al. "Mortality Due to schistosomiasis mansoni: A Field Study in Sudan." *American Journal of Tropical Medicine and Hygiene* 60, no. 2 (1999): 307–10.

Kinney, P. L. "Climate Change, Air Quality, and Human Health." *American Journal of Preventive Medicine* 35 (2008): 450–67.

Kirkpatric, B., L. E. Fleming, L. C. Backer, J. A. Bean, R. Tamer, G. Kirkpatrick, T. Kane et al. "Environmental Exposures to Florida Red Tides: Effects on Emergency Room Respiratory Diagnoses Admissions." *Harmful Algae* 5 (2006): 526–33.

Kloos, H. "Health Aspects of Resettlement in Ethiopia." *Social Science & Medicine* 30, no. 6 (1990): 643–56.

Knowlton, K. "Ozone, Oppressive Air Masses, and Degraded Air Quality." In *Global Climate Change and Human Health*, edited by G. Luber and J. Lemery, 138–70. San Francisco: John Wiley & Sons, 2015.

Kotloff, K. L., J. P Winickoff, B. Ivanoff et al. "Global Burden of Shigella Infections: Implications for Vaccine Development and Implementation of Control Strategies." World Health Organization, http://www.who.int/bulletin/archives/77(8)651.pdf, 1999. Accessed May 6, 2016.

Krishnamachari, K. A. V. R., V. Nagaraja, R. Bhat, and T. B. G. Tilak. "Hepatitis Due to aflatoxicosis: An Outbreak in Western India." *Lancet* 305, no. 7915 (1975): 1061–63.

Kuntz, J., and R. Murray. "Predictability of Swimming Prohibitions by Observational Parameters: A Proactive Public Health Policy, Stamford, Connecticut, 1989–2004." *Journal of Environmental Health* 72, no. 1 (2009): 17–22.

Lake Michigan Management Plan 2000, https://www.epa.gov/sites/production/files/2015-11/documents/lake-michigan-lamp-2000-458pp.pdf, April 2000. Accessed June 4, 2016.

Lau, C. L., L. D. Smythe, S. B. Craig, and P. Weinstein. "Climate Change, Flooding, Urbanisation and Leptospirosis: Fuelling the Fire?" *Transactions of the Royal Society of Tropical Medicine and Hygiene* 104, no. 10 (2010): 631–38.

Lavelle, M. "Has Climate Change Made Lyme Disease Worse?" *Scientific American*, https://www.scientificamerican.com/article/has-climate-change-made-lyme-disease-worse/. Accessed October 12, 2016.

Lazensky, B. "Florida Department of Health, Nassau County Health Department, Investigation of a Cluster of Ciguatera Fish Poisoning Cases (n = 13) in Restaurant Patrons Who Consumed Grouper." 2008, Unpublished data.

Lee, B. J., B. Kim, and K. Lee. "Air Pollution Exposure and Cardiovascular Disease." *Toxicol Res* 30, no. 2 (2014): 71–75.

Lemery, Jay. "Lessons From Dr. Strangelove." *Wilderness & Environmental Medicine* 23, no. 1 (2012): 2–4.

Lemery, J., C. Williams, and P. Farmer. "Editorial: The Great Procrastination." *Health & Human Rights: An International Journal* 16, no. 1 (2014): 1–3.

Lindblade, K. A., E. D. Walker, A. W. Onapa, J. Katungu, and M. L. Wilson. "Highland Malaria in Uganda: Prospective Analysis of an Epidemic Associated with El Niño." *Transactions of the Royal Society of Tropical Medicine and Hygiene* 93, no. 5 (1999): 480–87.

Liu, L., H. L. Johnson, S. Cousens, J. Perin, S. Scott, J. E. Lawn, I. Rudan, H. Campbell, R. Cibulskis, M. Li, C. Mathers, and R. E. Black, Child Health Epidemiology Reference Group of WHO and UNICEF. "Global, Regional, and National Causes of Child Mortality: An Updated Systematic Analysis for 2010 with Time Trends since 2000." *Lancet* 379, no. 9832 (2012): 2151–61.

Liu, S. K., S. Cai, Y. Chen et al. "The Effect of Pollutional Haze on Pulmonary Function." *J Thorac Dis* 8, no. 1 (2016): E41–56.

Lloyd, S. J., R. S. Kovats, and Z. Chalabi. "Climate Change, Crop Yields, and Undernutrition: Development of a Model to Quantify the Impact of Climate Scenarios on Child Undernutrition." *Environmental Health Perspectives* 119 (2011): 1817.

Lo, E., and E. Levetin. "Influence of Meteorological Conditions on Early Spring Pollen in the Tulsa Atmosphere from 1987–2006." *Journal of Allergy and Clinical Immunology* 119, no. 1S (2007): 101.

Loiseau, C., R. J. Harrigan, A. J. Cornel, S. L. Guers, M. Dodge, T. Marzec, . . . and R. N. Sehgal. "First Evidence and Predictions of Plasmodium Transmission in Alaskan Bird Populations." *PLoS One* 7, no. 9 (2012): e44729.

Loladze, I. "Hidden Shift of the Inome of Plants Exposed to Elevated CO2 Depletes Minerals at the Base of Human Nutrition." *eLife* 3 (2014): e02245. doi:107554/Elife.02245.

"Malé Declaration on the Human Dimension of Global Climate Change," http://www.ciel.org/Publications/Male_Declaration_Nov07.pdf, November 14, 2007. Accessed October 5, 2016.

Malilay, J. "Floods." In *The Public Health Consequences of Disasters*, edited by E. Noji, 287–301. New York: Oxford University Press, 1997.

Manuel, J. "In Katrina's Wake." *Environmental Health Perspective* 114, no. 1 (2006): A32–39.

Mas-Coma, S., M. A. Valero, and M. D. Bargues. "Climate Change Effects on Trematodiases, with Emphasis on Zoonotic Fascioliasis and Schistosomiasis." *Veterinary Parasitology* 163, no. 4 (2009): 264–80.

Mathers, C., D. M. Fat, and J. Boerma. *The Global Burden of Disease: 2004, Update.* Geneva: World Health Organization, 2008.

McCarthy, J. E., and R. K. Lattanzio. "Ozone Air Quality Standards: EPA's 2015 Revision." Congressional Research Service, https://fas.org/sgp/crs/misc/R43092.pdf, January 25, 2016. Accessed October 8, 2016.

McGranahan, Gordon, Deborah Balk, and Bridget Anderson. "The Rising Tide: Assessing the Risks of Climate Change and Human Settlements in Low Elevation Coastal Zones." *Environment and Urbanization* 19, no. 1 (2007): 17–37.

McLaughlin, J. B., A. DePaola, C. A. Bopp, K. A. Martinek, N. P. Napolilli, C. G. Allison, . . . and J. P. Middaugh. "Outbreak of Vibrio parahaemolyticus Gastroenteritis Associated with Alaskan Oysters." *New England Journal of Medicine* 353, no. 14 (2005): 1463–70.

McNeil, D. "For Intrigue, Malaria Drug Gets the Prize." *New York Times*, January 16, 2012, http://www.nytimes.com/2012/01/17/health/for-intrigue-malaria-drug-artemisinin-gets-the-prize.html. Accessed November 28, 2016.

Melilo, J., and O. Sala. "Ecosystem Services." In *Sustaining Life: How Human Health Depends on Biodiversity.* New York: Oxford University Press, 2008.

Millar, C. I., R. D. Westfall, D. L. Delany, J. C. King, and L. J. Graumlich. "Response of Subalpine Conifers in the Sierra Nevada, California, USA, to 20th-Century Warming and Decadal Climate Variability." *Arctic, Antarctic, and Alpine Research* 36 (2004): 181–200. doi:10.1657/1523-0430(2004)036[0181:roscit]2.0.co;2.

Mills, J. N., and J. E. Childs. "Ecologic Studies of Rodent Reservoirs: Their Relevance for Human Health." *Emerging Infectious Diseases, Perspectives* 4, no. 4 (1998): 529–37.

Mills, J. N., B. A. Ellis, K. T. McKee Jr., G. E. Calderon, J. I. Maiztegui, G. O. Nelson, . . . and J. E. Childs. "A Longitudinal Study of Junin Virus Activity in the Rodent Reservoir of Argentine Hemorrhagic Fever." *American Journal of Tropical Medicine and Hygiene* 47, no. 6 (1992): 749–63.

Mlingi, N., N. H. Poulter, and H. Rosling. "An Outbreak of Acute Intoxications from Consumption of Insufficiently Processed Cassava in Tanzania." *Nutrition Research* 12, no. 6 (1992): 677–87.

Molfino, N. A., S. C. Wright, I. Katz, S. Tarlo, F. Silverman, P. A. McClean, . . . and M. Raizenne. "Effect of Low Concentrations of Ozone on Inhaled Allergen Responses in Asthmatic Subjects." *Lancet* 338, no. 8761) (1991): 199–203.

Montgomery, M. P., F. Kamel, T. M. Salana, M. C. R. Alavanja, and D. P. Sandler. "Incident Diabetes and Pesticide Exposure among Licensed Pesticide Applicators: Agricultural Health Study 1993–2003." *Amer J Epidemiol* 167 (2008): 1235.

Moore, Sue E., and Henry P. Huntington. "Arctic Marine Mammals and Climate Change: Impacts and Resilience." *Ecological Applications* 18, sp2 (2008).

Moore, S. K., N. J. Mantua, B. M. Hickey, and V. L. Trainer. "Recent Trends in Paralytic Shellfish Toxins in Puget Sound, Relationships to Climate, and Capacity for Prediction of Toxic Events." *Harmful Algae* 8 (2009): 463–77.

Moses, M. "Pesticide-Related Health Problems and Farmworkers." *Official Journal of the American Association of Occupational Health Nurses* 37, no. 3 (1989): 115–30.

Muriel, P., T. Downing, M. Hulme, R. Harrington, D. Lawlor, D. Wurr, C. J. Atkinson et al. *Climate Change and Agriculture in the United Kingdom.* London: Ministry of Agriculture, Fisheries and Forestry, 2001.

National Aeronautics and Space Administration (NASA). "NASA Finds Drought in Eastern Mediterranean Worst of Past 900 Years," http://www.nasa.gov/feature/goddard/2016/nasa-finds-drought-in-eastern-mediterranean-worst-of-past-900-years, March 1, 2016. Accessed October 8, 2016.

———. http://www.nasa.gov/feature/goddard/2016/climate-trends-continue-to-break-records. Accessed October 12, 2016.

National Centers for Coastal Ocean Science. "How Climate Change Could Impact Harmful Algal Blooms," https://coastalscience.noaa.gov/news/climate/climate-change-impact-harmful-algal-blooms/, November 6, 2014. Accessed October 5, 2016.

National Centers for Environmental Information. https://www.ncdc.noaa.gov/sotc/national/201503. Accessed November 28, 2016.

National Hurricane Center. http://www.nhc.noaa.gov/data/tcr/EP202015_Patricia.pdf. Accessed October 12, 2016.

National Ice Core Laboratory. http://www.icecores.org/about/index.shtm. Accessed October 4, 2016.

National Interagency Fire Center (NIFC). *Wildland Fire Summary and Statistics Annual Report 2011.* Boise: National Interagency Fire Center, 2012.

National Ocean Service, "What Is a Red Tide?" http://oceanservice.noaa.gov/facts/redtide.html. Accessed October 5, 2016.

National Oceanic and Atmospheric Administration (NOAA). Centers for Environmental Information, https://www.ncdc.noaa.gov/data-access/paleoclimatology-data/datasets/ice-core. Accessed June 17, 2016.

———. https://www.climate.gov/news-features/understanding-climate/2013-state-climate-record-breaking-super-typhoon-haiyan. Accessed October 12, 2016.

———. http://research.noaa.gov/News/NewsArchive/LatestNews/TabId/684/ArtMID/1768/ArticleID/11153/Greenhouse-gas-benchmark-reached-.aspx. Accessed October 12, 2016.

National Weather Service. http://www.nws.noaa.gov/om/heat/heat_index.shtml. Accessed October 4, 2016.

———. http://www.nws.noaa.gov/om/hazstats.shtml. Accessed October 12, 2016.

Nelson, G. C. *Climate Change: Impact on Agriculture and Costs of Adaptation.* Washington, DC: International Food Policy Research Institute, 2009.

Nichols, R. W., and H. H. Schmitt. "The Phony War against CO2." *Wall Street Journal,* November 1, 2016, A11.

O'Connell, E. J. "The Burden of Atopy and Asthma in Children." *Allergy* 59, suppl 78 (2004): 7–11.

"Ocean Acidification: The Other Carbon Dioxide Problem," NOAA, http://www.pmel.noaa.gov/co2/story/Ocean+Acidification. Accessed June 4, 2016.

Ogden, N. H., A. Maarouf, I. K. Barker, M. Bigras-Poulin, L. R. Lindsay, M. G. Morshed, . . . and D. F. Charron. "Climate Change and the Potential for Range Expansion of the Lyme Disease Vector Ixodes scapularis in Canada." *International Journal for Parasitology* 36, no. 1 (2006): 63–70.

Olson, S. H., R. Gangnon, E. Elguero, L. Durieux, J. F. Guégan, J. A. Foley, and J. A. Patz. "Links between Climate, Malaria, and Wetlands in the Amazon Basin." *Emerg Infect Dis* 15, no. 4 (2009): 659–62.

Paaijmans, K. P., A. F. Read, and M. B. Thomas. "Understanding the Link between Malaria Risk and Climate." *Proceedings of the National Academy of Sciences* 106, no. 33 (2009): 13844–49.

Pablo, J., S. A. Banack, P. A. Cox, T. E. Johnson, S. Papapetropoulos, W. G. Bradley, . . . and D. C. Mash. "Cyanobacterial neurotoxin BMAA in ALS and Alzheimer's Disease." *Acta Neurologica Scandinavica* 120, no. 4 (2009): 216–25.

Parry, M., A. Evans, M. W. Rosegrant, and T. Wheeler. *Climate Change and Hunger: Responding to the Challenge.* Washington, DC: International Food Policy Research Institute, 2009.

Patra, S., A. Kumar, S. S. Trivedi, M. Puri, and S. K. Sarin. "Maternal and Fetal Outcomes in Pregnant Women with Acute Hepatitis E Virus Infection." *Ann Intern Med* 147 (2007): 28033. Accessed May 6, 2016.

Pearce, N., N. Ait-Khaled, R. Besley, J. Malloi, U. Keli, E. Mitchell, and C. Robertson. "Worldwide Trends in the Prevalence of Asthma Symptoms: Phase III of the International Study of Asthma and Allergies in Childhood (ISAAC)." *Thorax* 62 (2007): 758–66.

Perl, T. M., L. Bédard, T. Kosatsky, J. C. Hockin, E. C. Todd, and R. S. Remis. "An Outbreak of Toxic Encephalopathy Caused by Eating Mussels Contaminated with Domoic Acid." *New England Journal of Medicine* 322, no. 25 (1990): 1775–80.

Peters, A., H. E. Wichmann, T. Tuch, J. Heinrich, and J. Heyder. "Respiratory Effects Are Associated with the Number of Ultrafine Particles." *American Journal of Respiratory and Critical Care Medicine* 155, no. 4 (1997): 1376–83.

Peters, D. P. C., A. E. Lugo, F. S. Chapin III, S. T. A. Pickett, M. Duniway, A. V. Rocha, F. J. Swanson, C. Laney, and J. Jones. "Cross-System Comparisons Elucidate Disturbance Complexities and Generalities." *Ecosphere* 2 (2011): 1–26. doi:10.1890/ES11-00115.1.

Pfeffercorn, Elmer, author (JML) personal reference, https://geiselmed.dartmouth.edu/faculty/facultydb/view.php?uid=1525. Accessed October 2, 2016.

Philipsborn, R., S. M. Ahmed, B. J. Brosi, and K. Levy. "Climatic Drivers of Diarrheagenic Escherichia coli Incidence: A Systematic Review and Meta-analysis." *Journal of Infectious Diseases* (2016): jiw081.

Pimentel, D. "Climate Changes and Food Supply." *Forum for Applied Research and Public Policy* 8, no. 4 (1993): 54–60.

Poliomyelitis Pinkbook. Centers for Disease Control, http://www.cdc.gov/vaccines/pubs/pinkbook/downloads/polio.pdf, April 2015. Accessed May 6, 2016.

Pope, C. A. III, R. T. Burnett, M. J. Thun, E. E. Calle, and G. D. Thurston. 2002. "Lung Cancer, Cardiopulmonary Mortality and Long Term Exposure to Fine Particulate Air Pollution." *JAMA* 287: 1132–41.

Porter, J. R., L. Xie, A. J. Challinor, K. Cochrane, S. M. Howden, M. M. Iqbal, D. B. Lobell et al. "Food Security and Food Production Systems." In *Climate Change 2014: Impacts, Adaptation, and Vulnerability. Contribution of Working Group II to the Fifth Assessment Report of the Intergovernmental Panel on Climate Change*, edited by C. B. Field, V. R. Barros, D. Dokken, K. J. Mach, M. D. Mastrandrea, T. E. Bilir, M. Chatterjee et al., 485–533. Cambridge: Cambridge University Press, 2014.

Potts, S. G., J. C. Biesmeijer, C. Kremen, P. Neumann, O. Schweiger, and W. E. Kunin. "Global Pollinator Declines: Trends, Impacts and Drivers." *Trends in Ecology & Evolution* 25, no. 6 (2010): 345–53.

"Public Health Statement for DDT, DDE, and DDD." Agency for Toxic Substance and Disease Registry, Centers for Disease Control, http://www.atsdr.cdc.gov/phs/phs.asp?id=79&tid=20, September 2002. Accessed June 4, 2016.

"Quest Diagnostics Health Trends, Allergy Report 2011. Allergies Across America," https://www.questdiagnostics.com/dms/Documents/Other/2011_QD_AllergyReport.pdf, May 2011. Accessed September 26, 2016.

Raffa, K. F., B. H. Aukema, B. J. Bentz, A. L. Carroll, J. A. Hicke, M. G. Turner, and W. H. Romme. "Cross-scale Drivers of Natural Disturbances Prone to Anthropogenic Amplification: The Dynamics of Bark Beetle Eruptions." *BioScience* 58 (2008): 501–17. doi:10.1641/b580607.

Raleigh, C. "The Search for Safety: The Effects of Conflict, Poverty and Ecological Influences on Migration in the Developing World." *Global Environmental Change* 21 (2011): S82–S93.

Rappold, A. G., S. L. Stone, W. E. Cascio, L. M. Neas, V. J. Kilaru, M. S. Carraway, . . . and H. Vaughan-Batten. "Peat Bog Wildfire Smoke Exposure in Rural North Carolina Is Associated with Cardiopulmonary Emergency Department Visits Assessed through Syndromic Surveillance." *Environmental Health Perspectives* 119, no. 10 (2011): 1415.

Ratard, R., C. M. Brown, J. Ferdinands, and D. Callahan. "Health Concerns Associated with Mold in Water-Damaged Homes after Hurricanes Katrina and Rita— New Orleans, Louisiana, October 2005." *Morb Mortal Wkly Rep* 55, no. 22006 (March 10, 2006): 41–44.

Raymond, P. A., M. B. David, and J. E. Saiers. "The Impact of Fertilization and Hydrology on Nitrate Fluxes from Mississippi Watersheds." *Current Opinion in Environmental Sustainability* 4 (2012): 212–18. doi:10.1016/j.cosust.2012.04.001.

Raymond, P. A., N.-H. Oh, E. R. Turner, and W. Broussard. "Anthropogenically Enhanced Fluxes of Water and Carbon from the Mississippi River." *Nature* 451 (2008): 449–52. doi:10.1038/nature06505.

Regional Health Forum WHO South-East Asia Region 12, no. 1 (2008). "Special Issue on World Health Day 2008 Theme: Protecting Health from Climate Change," http://www.who.int/world-health-day/toolkit/report_web.pdf. Accessed March 16, 2017.

Reichwaldt, E. S., and A. Ghadouani. "Effects of Rainfall Patterns on Toxic Cyanobacterial Blooms in a Changing Climate: Between Simplistic Scenarios and Complex Dynamics." *Water Research* 46, no. 5 (2012): 1372–93.

Reid, Colleen E., and Janet L. Gamble. "Aeroallergens, Allergic Disease, and Climate Change: Impacts and Adaptation." *Ecohealth* 6, no. 3 (2009): 458–70.

Richer, R., S. A. Banack, J. S. Metcalf, and P. A. Cox. "The Persistence of Cyanobacterial Toxins in Desert Soils." *Journal of Arid Environments* 112, no. B (2015): 134–39.

Roberts, J. R., C. J. Karr, J. A. Paulson, A. C. Brock-Utne, H. L. Brumberg, C. C. Campbell, . . . and R. O. Wright. "Pesticide Exposure in Children." *Pediatrics* 130, no. 6 (2012): e1765–88.

Robinson, M. "Foreword." *Health and Human Rights Journal* 16, no. 1 (2014): 4–7.

Rodrigue, D. C., R. A. Etzel, S. Hall, E. De Porras, O. H. Velasquez, R. V. Tauze, . . . and P. A. Blake. "Lethal Paralytic Shellfish Poisoning in Guatemala." *American Journal of Tropical Medicine and Hygiene* 42, no. 3 (1990): 267–71.

Rojstaczer, S., S. M. Sterling, and N. J. Moore. "Human Appropriation of Photosynthesis Products." *Science* 294, no. 5551 (2001): 2549–52.

Roy, S. B., L. Chen, E. H. Girvetz, E. P. Maurer, W. B. Mills, and T. M. Grieb. "Projecting Water Withdrawal and Supply for Future Decades in the U.S. under Climate Change Scenarios." *Environmental Science & Technology* 46 (2012): 2545−56. doi:10.1021/es2030774.

Ruiz, M. O., L. F. Chaves, G. L. Hamer, T. Sun, W. M. Brown, E. D. Walker, . . . and U. D. Kitron. "Local Impact of Temperature and Precipitation on West Nile Virus Infection in Culex Species Mosquitoes in Northeast Illinois, USA." *Parasites & Vectors* 3, no. 1 (2010): 1.

Russac, P. A. "Epidemiological Surveillance: Malaria Epidemic Following the Nino Phenomenon." *Disasters* 10 (1986): 112–17.

Sabo, J. L., T. Sinha, L. C. Bowling, G. H. W. Schoups, W. W. Wallender, M. E. Campana, K. A. Cherkauer, P. L. Fuller, W. L. Graf, J. W. Hopmans, J. S. Kominoski, C. Taylor, S. W. Trimble, R. H. Webb, and E. E. Wohl. "Reclaiming Freshwater Sustainability in the Cadillac Desert." *Proceedings of the National Academy of Sciences*, 107, 2010: 21263–69. doi:10.1073/pnas.1009734108.

Salvi, R. M., D. R. Lara, E. S. Ghisolfi, L. V. Portela, R. D. Dias, and D. O. Souza. "Neuropsychiatric Evaluation in Subjects Chronically Exposed to Organophosphate Pesticides." *Toxicological Sciences* 7, no. 2 (2003): 267–71.

Scallan, E., R. M. Hoekstra, F. J. Angulo et al. "Foodborne Illness Acquired in the United States—Major Pathogens." *Emerg Infect Dis* 17 (2011): 7–15.

"Schistosomiasis Fact Sheet." World Health Organization, http://www.who.int/ schistosomiasis/epidemiology/en/, 2016. Accessed May 7, 2016.

Schmidhuber, J., and F. N. Tubiello. "Global Food Security under Climate Change." *Proceedings of the National Academy of Sciences* 104, no. 50 (2007): 19703–8.

Schoenwetter, W. F. "Allergic Rhinitis: Epidemiology and Natural History." *Allergy Asthma Proc* 21, no. 1 (2000): 1–6.

Schwartz, E. "Schistosomiasis." In *Tropical Diseases in Travelers*, edited by E. Schwartz. San Francisco: John Wiley & Sons, 2009.

Schwartz, J. "Lung Function and Chronic Exposure to Air Pollution: A Cross-Sectional Analysis of NHANES II." *Environmental Research* 50, no. 2 (1989): 309–21.

———. "Short Term Fluctuations in Air Pollution and Hospital Admissions of the Elderly for Respiratory Disease." *Thorax* 50, no. 5 (1995): 531–38.

Schwartz, J., D. Slater, T. V. Larson, W. E. Pierson, and J. Q. Koenig. "Particulate Air Pollution and Hospital Emergency Room Visits for Asthma in Seattle." *American Review of Respiratory Disease* 147, no. 4 (1993): 826–31.

Sedas, V. T. P. "Influence of Environmental Factors on the Presence of Vibrio cholerae in the Marine Environment: A Climate Link." *Journal of Infection in Developing Countries* 1, no. 3 (2007): 224–41.

Sellner, K. G., G. J. Doucette, and G. J. Kirkpatric. "Harmful Algal Blooms: Causes, Impacts and Detection." *Journal of Industrial Microbiology and Biotechnology* 30 (2003): 383–406.

Semenza, J. "Changes in Hydrology and Its Impacts on Waterborne Disease." In *Global Climate Change and Human Health*, edited by G. Luber and J. Lemery, 103–35. San Francisco: John Wiley & Sons, 2015.

Shea, K. M., R. T. Truckner, R. W. Weber, and D. B. Peden. "Climate Change and Allergic Disease." *Journal of Allergy and Clinical Immunology* 122 (2008): 443–53.

Sheffield, P. E., K. Knowlton, J. L. Carr, and P. L. Kinney. "Modeling of Regional Climate Change Effects on Ground-Level Ozone and Childhood Asthma." *American Journal of Preventive Medicine* 41 (2011): 251–57.

Shepard, C., V. N. Agostini, B. Gilmer, T. Allen, J. Stone, W. Brooks, and M. W. Beck. "Assessing Future Risk: Quantifying the Effects of Sea Level Rise on Storm Surge Risk for the Southern Shores of Long Island, New York." *Natural Hazards* 60 (2012): 727–45. doi:10.1007/s11069-011-0046-8.

Shukla, J. "Extreme Weather Events and Mental Health: Tackling the Psychosocial Challenge." *ISRN Public Health* (2013).

Silverstein, Daniel. "Food Security." Foundations for Global Health Responders Massive Open Online Course (Coursera, 2015), https://www.coursera.org/learn/ ghresponder.

Smith, K. "Climate Change, Violence, and the Afterlife. Health of People, Places and Planet," http://press-files.anu.edu.au/downloads/press/p320071/html/ch34. xhtml?referer=235&page=62, July 2015. Accessed October 5, 2016.

"Soil-Transmitted Helminth Infections Fact Sheet," World Health Organization, http:// www.who.int/mediacentre/factsheets/fs366/en/, March 2016. Accessed May 6, 2016.

Soper, G. A. "The Curious Career of Typhoid Mary." *Bulletin of the New York Academy of Medicine* 15, no. 10 (1939): 698.

Spiewak, R. "Pesticides as a Cause of Occupational Skin Diseases in Farmers." *Ann Agric Environ Med* 8, no. 1 (2001): 1–5.

Squires, S. "The World's Most Popular Pill Turns 100," *Washington Post*, August 5, 1997, https://www.washingtonpost.com/archive/lifestyle/wellness/1997/08/05/aspirin-the-worlds-most-popular-pill-turns-100/caa961d1-c7a4-42c7-b1ac-550193a9a21f/. Accessed October 12, 2016.

Stain, H., B. Kelly, T. Lewin, N. Higginbotham, J. Beard, and F. Hourihan. "Social Networks and Mental Health among a Farming Population." *Social Psychiatry and Psychiatric Epidemiology* 43 (2008): 843–49.

Stanke, C., M. Kerac, C. Prudhomme, J. Medlock, and V. Murray. "Health Effects of Drought: A Systematic Review of the Evidence." *PLOS Currents Disasters* (2013).

Stephenson, L. *The Impact of Helminth Infections on Human Nutrition.* London: Taylor and Francis, 1987.

Stone, Edmund. "An Account of the Success of the Bark of the Willow in the Cure of Agues. In a Letter to the Right Honourable George Earl of Macclesfield, President of RS from the Rev. Mr. Edmund Stone, of Chipping-Norton in Oxfordshire." *Philosophical Transactions* 53 (1763): 195–200.

Stott, P. A., D. A. Stone, and M. R. Allen. "Human Contribution to the European Heatwave of 2003." *Nature* 432, no. 7017 (2004): 610–14.

Sullivan, G., J. J. Vasterling, X. Han, A. T. Tharp, T. Davis, E. A. Deitch, and J. I. Constans. "Preexisting Mental Illness and Risk for Developing a New Disorder after Hurricane Katrina." *Journal of Nervous and Mental Disease* 201 (2013): 161–66. doi:10.1097/NMD.0b013e31827f636d.

Sunyer, J., C. Spix, P. Quenel, A. Ponce-de-Leon, A. Pönka, T. Barumandzadeh, . . . and L. Bisanti. "Urban Air Pollution and Emergency Admissions for Asthma in Four European Cities: The APHEA Project." *Thorax* 52, no. 9 (1997): 760–65.

Sykes, P. "Sinking States; Climate Change and the Next Refugee Crisis," https://www.foreignaffairs.com/articles/2015-09-28/sinking-states, September 28, 2015. Accessed October 8, 2016.

Tai, A. P., L. J. Mickley, and D. J. Jacob. "Correlations between Fine Particulate Matter (M 2.5) and Meteorological Variables in the United States: Implications for the Sensitivity of PM 2.5 to Climate Change." *Atmospheric Environment* 44, no. 32 (2010): 3976–84.

"Taking Action, Nutrition for Survival, Growth and Development," World Health Organization, http://www.who.int/pmnch/topics/child/acf_whitepaper.pdf, May 2010. Accessed June 4, 2016.

Tanner, T. L. "Rhus (Toxicodendron) Dermatitis." *Primary Care* 27, no. 2 (2000): 493–502.

Taub, D. R., B. Miller, and H. Allen. "Effects of Elevated Carbon Dioxide on the Protein Concentration of Food Crops: A Meta-Analysis." *Global Change Biology* (2008).

Tauxe, R. V., S. D. Holmberg, A. Dodin, J. V. Wells, and P. A. Blake. "Epidemic Cholera in Mali: High Mortality and Multiple Routes of Transmission in a Famine Area." *Epidemiology & Infection* 100, no. 2 (1988): 279–89.

"The State of Clinical Research in the United States: An Overview." In *Transforming Clinical Research in the United States: Challenges and Opportunities: Workshop Summary.* Institute of Medicine (U.S.) Forum on Drug Discovery, Development, and Translation. Washington, DC: National Academies Press, 2010.

Theobald, David M., and William H. Romme. "Expansion of the US Wildland–Urban Interface." *Landscape and Urban Planning* 83, no. 4 (2007): 340–54.

Tirado, M. C., P. Crahay, L. Mahy et al. "Climate Change and Nutrition: Creating a Climate for Nutrition Security." *Food and Nutrition Bulletin* 34, no. 4 (2013).

Turner, R. B., and R. B. Woodward. "The Chemistry of the Cinchona Alkaloids." *Alkaloids: Chemistry and Physiology* 3 (1953): 1–63.

Tylleskär, T. "The Association between Cassava and the Paralytic Disease Konzo." *Acta Hortic* 375 (1994): 331–40. doi:10.17660/ActaHortic.1994.375.33.

UK Health Alliance on Climate Change. http://www.ukhealthalliance.org/new-report-breath-fresh-air-addressing-climate-change-air-pollution-together-health/. Accessed November 28, 2016.

UK Royal Society. *Ocean Acidification Due to Increasing Atmospheric Carbon Dioxide.* Cardiff: Clyvedone Press, 2005.

Umlauf, G., G. Bidoglio, E. H. Christoph, J. Kampheus, F. Krüger, D. Landmann, . . . and D. Stehr. "The Situation of PCDD/Fs and Dioxin-like PCBs after the Flooding of River Elbe and Mulde in 2002." *Acta Hydrochimica et Hydrobiologica* 33, no. 5 (2005): 543–54.

United Nations (UN), Committee on Economic, Social and Cultural Rights. "General Comment 14," http://www.ohchr.org/Documents/Issues/Women/WRGS/Health/GC14.pdf, August 11, 2000. Accessed October 5, 2016.

———. "Development Program: Multidimensional Poverty Index," http://hdr.undp.org/en/content/multidimensional-poverty-index-mpi. Accessed November 28, 2016.

———. Educational, Scientific and Cultural Organization (UNESCO). "Securing the Food Supply 2001a," Paris, http://webworld.unesco.org/water/wwap/facts_figures/food_supply.shtml. Accessed October 12, 2016.

———. http://www.un.org/en/development/desa/news/population/world-urbanization-prospects-2014.html. Accessed October 4, 2016.

———. "Human Rights and Climate Change," http://www.ohchr.org/Documents/Issues/ClimateChange/A.HRC.RES.18.22.pdf?. Accessed March 16, 2017.

———. Office for Disaster Risk Reduction (UNISDR). "The Human Cost of Weather-Related Disasters 1995–2015," https://www.unisdr.org/2015/docs/climatechange/COP21_WeatherDisastersReport_2015_FINAL.pdf. Accessed March 16, 2017.

———. "Universal Declaration of Human Rights," http://www.un.org/en/universal-declaration-human-rights/, December 10, 1948. Accessed October 5, 2016.

———. "Water Thematic Paper. Transboundary Waters: Sharing Benefits, Sharing Responsibilities," http://www.unwater.org/downloads/UNW_TRANSBOUNDARY.pdf, 2008. Accessed October 12, 2016.

———. World Health Organization. Mental Health Evidence, Research Team, and Disease Control Priorities Project. Disease Control Priorities Related to Mental, Neurological, Developmental and Substance Abuse Disorders, 2006.

———. "World Health Organization, Europe: Health Risks of Particulate Matter from Long-Range Transboundary Air Pollution," http://www.euro.who.int/__data/assets/pdf_file/0006/78657/E88189.pdf. Accessed March 16, 2017.

———. "World Population Aging Report," http://www.un.org/en/development/desa/population/publications/pdf/ageing/WPA2015_Report.pdf. Accessed October 4, 2016.

———. "World Urbanization Prospects," https://esa.un.org/unpd/wup/Publications/Files/WUP2014-Highlights.pdf. Accessed November 28, 2016.

———. "World Urbanization Prospects Report," http://www.un.org/en/development/desa/news/population/world-urbanization-prospects-2014.html. Accessed October 12, 2016.

———. "World Water Development Report," http://www.unwater.org/publications/world-water-development-report/en/. Accessed April 10, 2016. United Nations High Commissioner for Refugees. "Global Trends; Forced Displacement in 2015," http://www.unhcr.org/576408cd7.pdf, 2015. Accessed October 8, 2016.

United States Army Public Health Command. "Charmak Disease," https://phc.amedd.army.mil/PHC%20Resource%20Library/Charmak%20Jan%202010.pdf. Accessed October 5, 2016.

United States Department of Agriculture. "Foreign Agricultural Service GAIN Report," http://gain.fas.usda.gov/Recent%20GAIN%20Publications/Red%20Tide%20and%20Labor%20Unrest%20Reduce%20Chilean%20Salmon%20Production_Santiago_Chile_7-5-2016.pdf, July 5, 2016. Accessed November 14, 2016.

United States Environmental Protection Agency, https://www.epa.gov/climate-change-science/understanding-link-between-climate-change-and-extreme-weather. Accessed October 4, 2016.

United States Geological Survey, https://www2.usgs.gov/faq/node/3469. Accessed October 12, 2016.

———. "Toxic Substances Hydrology Program," http://toxics.usgs.gov/highlights/algal_toxins/algal_faq.html. Accessed October 12, 2016.

United States Global Change Research Program. "Review of the Impacts of Climate Variability and Change on Aeroallergens and Their Associated Effects," http://static1.1.sqspcdn.com/static/f/551504/6467325/1270769757893/GCRP.pdf?token=w6WDUSqpLQ6D%2B%2BljSLqMKAWExYQ%3D. Accessed March 16, 2017.

University of Arizona, UA News. "Effects of Climate Change on West Nile Virus," https://uanews.arizona.edu/story/effects-of-climate-change-on-west-nile-virus, September 9, 2013. Accessed October 8, 2016.

Veith, G. D. "Baseline Concentrations of Polychlorinated Biphenyls and DDT in Lake Michigan Fish, 1971." *Pestic Monit J* 9, no. 1 (June 1975): 21–29.

Vitousek, P. M., H. A. Mooney, J. Lubchenco, and J. M. Melillo. "Human Domination of Earth's Ecosystems." *Science* 277, no. 5325 (1997): 494–99.

Vos, T., A. D. Flaxman, M. Naghavi, R. Lozano, C. Michaud, M. Ezzati, K. Shibuya, J. A. Salomon et al. "Years Lived with Disability (Ylds) for 1160 Sequelae of 289 Diseases and Injuries 1990–2010: A Systematic Analysis for the Global Burden of Disease Study 2010." *Lancet* 380, no. 9859 (December 15, 2012): 2163–96.

Watkins, K. "Human Development Report," UN Development Programme 2007.

Watson, J. T., M. Gayer, and M. A. Connolly. "Epidemics after Natural Disasters." *Emerg Infect Dis* 13, no. 1 (2007): 1–5.

Watts, N., N. W. Adger, P. Agnolucci et al. "Health and Climate Change: Policy Responses to Protect Public Health." *Lancet* 386, no. 10006 (2015): 1861–1914.

Weber, R. W. "Patterns of Pollen Cross-Allergenicity." *Journal of Allergy and Clinical Immunology* 112, no. 2 (2003): 229–39.

Wellburn, A. "Atmospheric Nitrogenous Compounds and Ozone—Is NOx Fixation by Plants a Possible Solution?" *New Phytologist* 138, no. 1 (1998): 5–9.

Wenger, S. J., D. J. Isaak, C. H. Luce, H. M. Neville, K. D. Fausch, J. B. Dunham, D. C. Dauwalter, M. K. Young, M. M. Elsner, B. E. Rieman, A. F. Hamlet, and J. E. Williams. "Flow Regime, Temperature, and Biotic Interactions Drive Differential Declines of Trout Species under Climate Change." *Proceedings of the National Academy of Sciences* 108 (2011): 14175–80. doi:10.1073/pnas.1103097108.

Westerling, A. L., H. G. Hidalgo, D. R. Cayan, and T. W. Swetnam. "Warming and Earlier Spring Increase Western U.S. Forest Wildfire Activity." *Science* 313, no. 5789 (2006): 940–43.

Williams, C. "Climate Change and the Right to Health." In *Global Climate Change and Human Health*, edited by G. Luber and J. Lemery, 601–15. San Francisco: John Wiley & Sons, 2015.

Wisnivesky, J. P., S. L. Teitelbaum, A. C. Todd et al. "Persistence of Multiple Illnesses in World Trade Center Rescue and Recovery Workers: A Cohort Study." *Lancet* 378 (2011): 898–905.

Wong, T. W., T. S. Lau, T. S. Yu, A. Neller, S. L. Wong, W. Tam, and S. W. Pang. "Air Pollution and Hospital Admissions for Respiratory and Cardiovascular Diseases in Hong Kong." *Occupational and Environmental Medicine* 56, no. 10 (1999): 679–83.

World Bank. "Poverty," http://www.worldbank.org/en/topic/poverty. Accessed November 28, 2016.

World Health Organization (WHO). "Air Quality Guidelines: Global Update 2005: Particulate Matter, Ozone, Nitrogen Dioxide, and Sulfur Dioxide," 2006.

———. "Diarrhea," http://www.who.int/mediacentre/factsheets/fs330/en/. Accessed October 12, 2016.

———. "A Global Brief on Vector-Borne Diseases," http://apps.who.int/iris/bitstream/10665/111008/1/WHO_DCO_WHD_2014.1_eng.pdf, 2014. Accessed October 8, 2016. Document no. WHO/DCO/WHD/2014.1

———. http://www.who.int/gho/urban_health/situation_trends/urban_population_growth_text/en. Accessed October 12, 2016.

———. http://www.who.int/csr/don/2012_10_17/en/. Accessed October 4, 2016.

———. "Malaria," http://www.who.int/mediacentre/factsheets/fs094/en/. Accessed October 4, 2016.

———. "Zika," http://www.who.int/mediacentre/factsheets/zika/en/. Accessed October 4, 2016.

Yang, G. Q., S. Z. Wang, R. H. Zhou, and S. Z. Sun. "Endemic Selenium Intoxication of Humans in China." *American Journal of Clinical Nutrition* 37, no. 5 (1983): 872–81.

Yohe, G. W., R. D. Lasco, Q. K. Ahmand, N. W. Arnell, S. J. Cohen, C. Hope et al. "Perspectives on Climate Change and Sustainability." In *Climate Change 2007:*

Impacts, Adaptation and Vulnerability, Contribution of Working Group II to the Fourth Assessment Report of the Intergovernmental Panel on Climate Change, edited by M. L. Parry, O. F. Canziani, J. P. Palutikof, P. J. van der Linden, and C. E. Hanson, 811–41. Cambridge: Cambridge University Press, 2007.

Zahm, S. H., and A. Blair. "Pesticides and Non-Hodgkin's Lymphoma." *Cancer Res* 52, 19 suppl (October 1992): 5485–88.

Zanobetti, A., M. S. O'Neill, C. J. Gronlund, and J. D. Schwartz. "Summer Temperature Variability and Long-Term Survival among Elderly People with Chronic Disease." *Proceedings of the National Academy of Sciences of the United States of America* 109, no. 17 (2012): 6608–13. doi:10.1073/pnas.1113070109.

Zhou, X. N., G. J. Yang, K. Yang, X. H. Wang, Q. B. Hong, L. P. Sun, . . . and J. Utzinger. "Potential Impact of Climate Change on Schistosomiasis Transmission in China." *American Journal of Tropical Medicine and Hygiene* 78 no. 2 (2008): 188–94.

Zika Foundation, https://zikafoundation.org/. Accessed October 4, 2016.

Ziska, L. H., and K. L. Ebi. "Climate Change, Carbon Dioxide, and Public Health: The Plant Biology Perspective." In *Global Climate Change and Human Health*, edited by G. Luber and J. Lemery, 195–213. San Francisco: John Wiley & Sons, 2015.

Ziska, L. H., K. Knowlton, C. Rogers, D. Dalan, N. Tierney, M. A. Elder, W. Filley et al. "Recent Warming by Latitude Associated with Increased Length of Ragweed Pollen Season in Central North America." *Proceedings of the National Academy of Sciences USA* 108 (2001): 4248–51.

Index

About the Authors

Jay Lemery, MD, is associate professor of Emergency Medicine at the University of Colorado School of Medicine and is chief of the Section of Wilderness and Environmental Medicine. He is a past president of the Wilderness Medical Society.

Dr. Lemery has an academic expertise in austere and remote medical care as well as the effects of climate change on human health. He has provided medical direction to National Science Foundation researchers operating at both poles, most recently serving as the EMS medical director for the U.S. Antarctic Program. He currently is a consultant for the Climate and Health Program at the Centers for Disease Control and Prevention and sits on the National Academy of Medicine's Roundtable on Environmental Health Sciences, Research, and Medicine. In 2016, he and like-minded faculty launched the University of Colorado's Consortium on Climate Change & Health.

He currently holds an academic appointment at the Harvard T. H. Chan School of Public Health (FXB Center for Health and Human Rights), where he is a contributing editor for its journal, *Health and Human Rights*. He is a coeditor of the textbook *Global Climate Change & Human Health: From Science to Practice*.

Paul Auerbach, MD, is the Redlich Family Professor in the Department of Emergency Medicine at the Stanford University School of Medicine and adjunct professor of Military/Emergency Medicine at the F. Edward Hébert School of Medicine of the Uniformed Services University of the Health Sciences. He is a founder and past president of the Wilderness Medical Society and elected member of the Council on Foreign Relations. Dr. Auerbach is editor of the definitive textbook *Wilderness Medicine* and author of *Field Guide to Wilderness Medicine* and *Medicine for the Outdoors*. He was the founding coeditor of the journal *Wilderness & Environmental Medicine* and is one of the world's leading experts in wilderness medicine and emergency medicine. Dr. Auerbach served as a first responder to the earthquakes in Haiti (2010) and Nepal (2015) and was instrumental in creation of the Nepal Ambulance Service, among other volunteer efforts.

Dr. Auerbach was an early proponent of physicians becoming active participants in the discussions on issues related to the environment and global climate change through a commentary published in the *Journal of the American Medical Association* in 2008 titled "Physicians and the Environment" and creation of the Environmental Council of the Wilderness Medical Society. He has been honored by the Divers Alert Network as the DAN/Rolex Diver of the Year and with a NOGI Award for Science from the Academy of Underwater Arts and Sciences, and recognized by the 98th Civil Affairs Battalion (Airborne) for his work in Haiti. He continues to seek opportunities to make the world a better place.